# Weaving
## The Universe
### Is Modern Cosmology
### Discovered or Invented?

# *Weaving* The Universe

## Is Modern Cosmology Discovered or Invented?

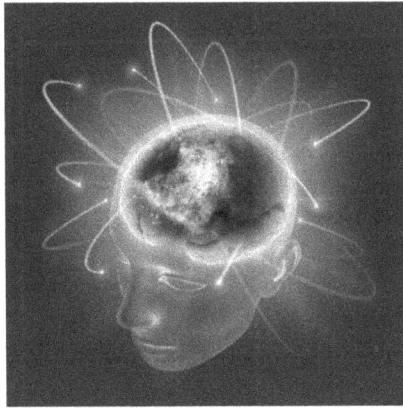

## Paul S. Wesson
University of Waterloo, Canada

**World Scientific**

NEW JERSEY · LONDON · SINGAPORE · BEIJING · SHANGHAI · HONG KONG · TAIPEI · CHENNAI

*Published by*

World Scientific Publishing Co. Pte. Ltd.

5 Toh Tuck Link, Singapore 596224

*USA office:* 27 Warren Street, Suite 401-402, Hackensack, NJ 07601

*UK office:* 57 Shelton Street, Covent Garden, London WC2H 9HE

**British Library Cataloguing-in-Publication Data**
A catalogue record for this book is available from the British Library.

**WEAVING THE UNIVERSE**
**Is Modern Cosmology Discovered or Invented?**

ISBN-13 978-981-4313-94-0
ISBN-10 981-4313-94-7

Printed by FuIsland Offset Printing (S) Pte Ltd. Singapore

# Other Books by Paul S. Wesson

Cosmology and Geophysics

Gravity, Particles and Astrophysics

Gravitation (with Robert B. Mann)

Space-Time-Matter

The Interstellar Undertakers

Dark Sky, Dark Matter (with James Overduin)

Cosmic Dreams

Brave New Universe (with Paul Halpern)

Five-Dimensional Physics

The Light/Dark Universe (with James Overduin)

Gambling with Galaxies

# PREFACE

The philosopher Descartes famously said "I think, therefore I am." The modern theoretical physicist might say "I think, and can invent the universe." That science is more subjective than commonly believed was put forward by the great astronomer Sir Arthur Eddington, who concluded that "the stuff of the world is mind-stuff." The aim of this book is to take a fresh look at the idea that physics is not so much discovered as invented, or that modern cosmology is studying what might be called the imagined universe.

I absorbed the works of Eddington while doing graduate work in cosmology at Cambridge in the 1970s. His views were controversial in the 1930s, but it seems to me that modern results in quantum theory and relativity have made them more acceptable. During a career spent largely solving problems having to do with the big bang, I have come to see that modern science has less to do with experiments than with the marvelous machinations of the human mind.

The cosmologist with his mathematics is rather like the weaver at his loom. The weaver sets up the basic lines or warp, sorts through bundles of material, and adds this as the weft to produce a garment. The scientist sets up his laws, considers the properties of matter, and if skillful enough puts out a coherent account of the universe. Though the analogy should not be pushed, both the weaver and the cosmologist are subject to judgment by those who follow fashion. I do not know if my views on the nature of science will fare any better than those of Eddington. In any case, no blame attaches to the colleagues who assisted my thoughts. These include the philosopher John Leslie (who came up with the mind-weaving analogy), the historian Martin Clutton-Brock, the cosmologist Paul Halpern and the physicist Francis Everitt (who also provided hospitality at Gravity Probe B). And of course there are other, more subtle acknowledgements, which follow from the thesis that theoretical physics is akin to other products of the human mind, like classical music, prose and fine art.

This book is directed toward those who think about things. Though it is written from the viewpoint of a professional cosmologist, I hope it will prove enlightening to anybody who takes "time out" to reflect on existence and enjoy the results of our culture, however it is expressed. (As I suggest in the text, the equation of the physicist can be viewed as a kind of miniature work of art, to be noted and admired before moving on.) Each person has an individual kind of mind syntax, which allows them to appreciate the achievements of other people in a unique way. Art, music and literature have generic forms of syntax, so they are appreciated and understood by large numbers of people. I am suggesting that science is comparable. Einstein's field equations of general relativity are just as rewarding to the physicist as the performance of a Tchaikovsky ballet is to the person who likes dance. In this book, I hope to broaden appreciation of something we all share, namely the power of the human mind.

Paul S. Wesson

# CONTENTS

# Chapter 1

# WEAVING THE WARP

The idea that science is at least partly invented, rather than discovered, was put forward most noticeably by the great astronomer Sir Arthur Eddington (1882–1944). He was severely criticized by both philosophers and physicists. However, recent advances in quantum mechanics and relativity have supported his thesis. In fact, it is now possible to present a fresh approach to the idea that science depends not so much on experiments as on the logical fit of theories coming from the human brain. It is the aim of this book to examine the ability of the human intellect to create science — or (in short) to study mind weaving.

Weaving in the traditional sense involves setting up on a loom the basic lines which determine structure (the warp), and adding to these the orthogonal threads which yield the colour and texture of the resulting fabric (the weft or woof). Modern science is like this, insofar as it involves basic laws, to which are added interpretations, resulting in an account of a specific part of the natural world. It is currently the aim of this scientific mind weaving to produce pieces of 'fabric', for example quantum mechanics and relativity theory; and to stitch these together to form a tapestry, or grand-unified theory of physics.

This is a laudable goal. But it is by no means obvious how to achieve it, or whether it is in principle achievable at all. It is traditional to separate physics — somewhat crudely — into the theoretical and experimental approaches. However, most physicists agree that the design, construction, and operation of an experiment involve theoretical elements; and certainly, the interpretation of the data from an experiment is mathematical and mind-based in nature. Eddington himself worked with observations in his former years, but later came to the view that physics (and science in general) is an intellectual exercise (Figure 1.1). We now have far more information at our disposal than did Eddington. So it is not surprising that some of our conclusions will differ from his. To present the modern argument for the mind as the seat of science, we have divided the material in a pragmatic fashion: Chapter 1 deals with the warp of scientific theory, while Chapter 7 deals with the weft of interpretation. The intervening Chapters 2–6 present the accepted elements of physics, though the presentation may be somewhat novel. This sandwich mode — philosophy in two slices of bread with the meat of physics between — is designed to present our arguments in the most efficient manner. Efficiency, at least in physics, is formalized by the philosophical statement called Ockham's razor. This is really an application of convenience or common sense insofar as it means that we introduce the least number of hypotheses necessary to solve a given problem. It is also widely used to choose between several viable theories for an observation, by taking the most simple.

**Figure 1.1.** Eddington, who was the Plumian Professor of Astronomy at Cambridge, came to believe that much of science is the product of the human mind.

A concept related to simplicity, which is much used in the quantitative sciences, is that of the minimum. We form a quantity which is typical of the system, and find the conditions under which it has its least value. The conditions found this way usually correspond to laws of nature. We need to understand this method before proceeding, and choose to illustrate it by two wide-ranging applications, one to the motion of a test particle and one to the laws which govern matter.

Measuring the distance between two points A and B in a given type of 'space' is arguably the most basic operation in physics, and was formalized by Euler, Fermat and others. On a flat, two-dimensional surface like the page of this book, there are an infinite number of paths connecting A and B. But one is special, namely that which makes the distance a minimum, giving a straight line (Figure 1.2). This is elementary; but already we see that a certain degree of subjectivity has entered our considerations, in that the concept of simplicity is based in the human mind. Particles which are not acted on by external forces travel on straight lines. It is worth consideration that physics would be unworkably complicated without this stricture. We may not, however, be able to measure the total distance between A and B, and only have access to a small element of it, say $ds$. Then we imagine that we can form the total distance, or interval, by integrating. If we vary the interval between A and B, keeping these points fixed, we can find the minimum. Technically, the mathematiccal problem involved here gives the extremum, but we conventionally disregard the maximum and choose the minimum (again, this is a subjective choice). The definition of a straight line then takes the symbolic form $\delta\left[\int ds\right] = 0$. This also gives the shortest (or 'straightest') path when the surface under consideration is not flat, but curved. There is also no restriction as to the number of dimensions of the 'space' involved, so the noted formula can be

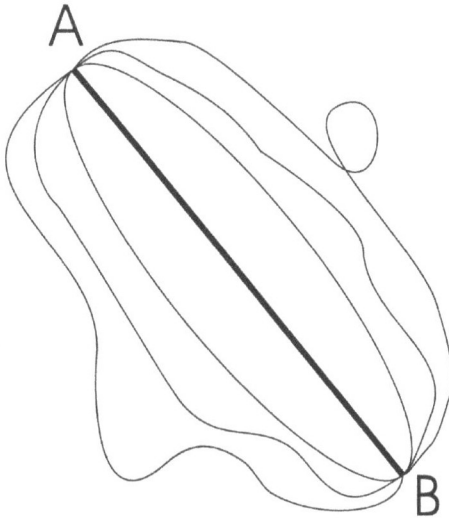

**Figure 1.2.** Between two points A and B there are an infinite number of curved lines, but a unique straight and shortest one.

applied to the four-dimensional space or manifold of relativity. The paths of particles travelling in the manifold are called geodesics. This word reminds us that on the curved 2D surface of the Earth, the geodesics are great circles, which define the most economical routes for travelling by airplanes when A and B denote cities. The crucial thing is that the Fermat principle can be applied to a manifold with any number of dimensions that is flat or curved, and serves to pick out a unique path from the infinitude that are possible. The principle can be applied in many situations, and was used especially to study the propagation of light, not only through empty, ordinary 3D space but also through the refractive 'space' provided by substances such as glass. In another context, sociologists have spent considerable effort trying to explain why most people value the truth above other forms of statement in everyday discourse. From the perspective of the Fermat principle, the answer is obvious: there are an infinite number of possible lies for a special and unique truth.

Matter is usually thought of as some kind of material which occupies a certain volume of ordinary 3D space and may evolve with time. In the above, we looked at the motion of a particle between two points in what we intrinsically assumed to be empty space. To handle the problem of matter, we could redo the analysis to take into account its effects on the motion of a test particle. However, we can actually go much farther than this, if we apply a more complicated version of the Fermat rule to the matter itself. That is, we can pick out rules for the matter's own behaviour by asking that it obey certain rules of minimality. The technique involved is called the calculus of variations. It was first used in the present context by Hilbert, who confirmed the importance of a quantity suspected as a 'catch-all' description of matter by Einstein.

General relativity is based on the insight by Einstein that the properties of matter in space and time are equivalent to the geometrical properties of 4D spacetime. This is a startling idea, beautiful in conception and successful in application. It is sometimes explained by the statement that matter curves spacetime, so by studying the mathematics of the latter we can work out the physics of the former. This statement is true as far as it goes. But what Einstein really showed was that matter and geometry are essentially the same thing. Ergo, if we wish to understand the laws of matter, we have to find a way to isolate preferred forms of the geometry.

This is where the calculus of variations comes in. There is a quantity in the geometry of curved spaces which is known as the Ricci scalar $R$. As a scalar, it is a simple thing, depending only on the location in space and time. (It lacks the directional properties of a vector, or the more complicated properties associated with the indices of a tensor.) Geometrically, it can be thought of as measuring the (inverse square of the) radius of curvature. Physically, it can be thought of roughly as measuring the energy density at a point in spacetime. Let the 'true' (corrected for curvature) volume element of a localized portion of spacetime be denoted $dV$. Then by analogy with our previous application of the Fermat rule, we can imagine that we integrate over the volume, take the variation, and set it to zero to get the minimum: $\delta\left[\int RdV\right] = 0$. The result identifies a quantity called the Einstein tensor, which is the basis of the gravitational field as it is described by general relativity.

The full theory, following Einstein, involves equating this geometrical tensor to the physical one which encodes familiar properties of matter such as the density and pressure. This material object is called the energy-momentum tensor (see Chapter 5 for a discussion of the properties of matter and tensors). In this way, general relativity gives a geometrical description of matter.

The approach outlined above can easily be extended to 'spaces' with more than the 4 dimensions of spacetime. In fact, there is nothing in the approach which restricts the number of dimensions of the manifold (see Chapter 3 for a discussion of the physical and mathematical aspects of dimensions). And we will see later that the extension of general relativity to $N = 5$ dimensions has the neat property of amalgamating the expressions for the $N = 4$ Einstein and energy-momentum tensors. By counting, the number of field equations for an ND theory of the type we are discussing is $N(N+1)/2$. These can be solved to obtain the allowed physics. (A more detailed account of the number and nature of ND field equations is given in Chapter 5.) In the 4D spacetime of general relativity, there are 10 relations, which are Einstein's field equations. Numerous exact solutions of these equations are known, and a compendium is due to Kramer et al. (1980). In the simplest extension of general relativity to the 5D manifold of space-time-matter theory and membrane theory, there are 15 relations, which express the most basic kind of unified field theory. Many exact solutions of these more difficult equations are known, and a compendium is due to Wesson (2007). At this stage, the reader may be feeling slightly stunned by the escalation involved in going from the motion of a test particle to the laws of gravitating matter. Take heart! We have, in a couple of pages, managed to write down a protocol for describing much of the physical world as it is currently understood. We have succeeded in reducing multiple infinities of possibility to a relatively few likelihoods.

The audacity of the human mind is truly remarkable. However, in contemplating the achievements of physics, we should not lose sight of the fact that its equations and associated paraphernalia are the manifestations of a kind of academic instinct. There is a parallel between the researcher in theoretical physics and the composer of classical music. The latter learns technique developed over several centuries, including the language of the stave and the technical properties of the musical instruments that make up an orchestra. Equipped with this learning, it is possible to write a symphony that is deep in technical quality, agreeable to the ear, and (hopefully) makes contact with those human feelings which are difficult to put into words but nonetheless vital. Connections between physics and music range from the incidental to the near profound (Halpern 2000, 2004). Einstein, of course, gained solace from playing the violin; and Feyman let out his energy by beating the bongo drums. Sir Fred Hoyle developed a theory of time by using classical music as a backdrop, and wrote a space-based opera. Sir James Jeans, who was a contemporary of Sir Arthur Eddington, even went so far as to propose that God must be both a mathematician and a musician. We will return to some of these topics later (see Chapters 4 and 5; also Halpern and Wesson 2006; Eddington 1928, 1939; and Hoyle 1966, 1994). Here we note that theoretical physics — like classical music or fine art — does not merely represent a job, but is a calling.

Cosmologists are an especially dedicated bunch. It is unknown how many professional cosmologists there are in the world, but they are probably no more numerous than brain surgeons. This in an age when popular coverage of the universe, particularly by television, gives the impression that it is easily understood. In this regard, it is instructive to look at some hard numbers. Today, a typical university will offer classes in astrophysics from first to fourth year which reveal a kind of pyramid structure. The enrolment in a first-year survey course on astronomy may typically be about 250. The following second-year class, which is frequently on the solar system, may have a student number of 70–100. By third year, when the subject is again restricted to a subfield such as stars or galaxies, the enrolment is down to about 30. The typical university will round out its educational offerings with a fourth-year course specifically on extragalactic astrophysics or cosmology, where it is fortunate if the attendance is 12. The student who survives the whole curriculum will indeed end up with a broad knowledge of the subject — which is after all what the word "cosmology" means. But the progressive cuts in enrolment, from hundreds to a dozen or so, tells us that the subject matter is not easy. This is partly because the dosage of mathematics increases as the educational process moves forward. In fact, the average television viewer of a show about the universe would likely be dismayed to discover that by the end of the study route for a B.Sc., the subject being taught is close to applied mathematics. The difference between astronomy and cosmology is akin to the difference between botany and genetics: one is mainly descriptive while the other is largely analytical.

Winnowing of the student population continues, moreover, at the postgraduate level. It typically takes two years in North America to complete an M.Sc. degree. And it is only towards the end of this (and then only for those at an academically strong institution) that the student has the opportunity to work on new material. However, the number of professional jobs in theoretical astrophysics or cosmology is so low that a minimum qualification for one is not an M.Sc. but a Ph.D. The latter is a particularly time-consuming project. It is not surprising that many gifted students abandon the quest at this stage, in favour of money/stability/family, rather than spend another segment of their life on something as esoteric as the big bang. There is a wide variation in the time that people spend on acquiring a doctoral degree. It lies for most able scholars in the range 4–8 years. Not only is this a significant chunk out of anybody's life; but it is also a period that for many is fraught with problems to do with research, arguments with supervisors and other academics, and the frustrating lack of cash. Wait a second, though. The course is not yet complete for the majority, even on completion of the Ph.D. degree. The paucity of jobs is such that most researchers will find themselves doing at least one post-doctoral two-year stint at a university, analogous to the internship at hospitals required of medical physicians. Enough! By age 32–35 typically, the scholar who is gifted enough and stubborn enough will finally obtain a position as a professional cosmologist.

The preceding account is conservative, number-wise. It deserves to be more widely appreciated than is apparently the case. Biblically, we are informed that "three score and ten" is about as much life as the average person can expect; and while modern medicine may enhance this somewhat, it is still true that most people's mental faculties are circumscribed by age 70. It is a sobering realization that for the typical cosmologist, half of his or her life is over before access to a regular pay cheque.

Why then do people aspire to become cosmologists? We can answer this superficially by repeating that it is not so much a job as a calling. However, a deeper insight can be gained by shifting the question to related fields. Why does the aspiring composer hope to emulate Beethoven, who when largely deaf managed to write his monumental ninth symphony? Or, going in another direction: Why does the young chess player try to emulate the brilliant gambits of the masters, like Bobby Fischer, Garry Kasparov and Boris Spassky? Above, we have remarked on the parallels between science and subjects in the arts, such as music. In fact, several leading cosmologists have likened their subject to some vast and intricate game of cosmic chess (Halpern and Wesson 2006). It is in a consideration of other subjects that we find an answer (at least partially) to the question of why some people are driven to study science and especially theoretical physics. Research in science means doing something new, and if it happens to have some relationship to the real world then so much the better.

Doing something new is usually satisfying — and even intoxicating — for the achiever. However, in science we have to be careful concerning what we mean by "new".

To the majority of scientists, doing something new means discovering an aspect of the natural world that was previously hidden from human appreciation, though the data are assumed to exist independent of the inquirer, who is like an explorer uncovering the plan of some concealed city of knowledge. This view is so traditional among scientists that we do not need to mention the names of those who have and still hold it.

To a few scientists who follow Sir Arthur Eddington, doing something new means using the power of the intellect to create fresh insights, whose development is mainly guided by the need for new knowledge to fit consistently with old and accepted knowledge. This view is rare, but puts science in the same class of human cultural achievements as (say) classical music and fine art.

In the last chapter of this book, we will argue in favour of the second opinion over the first. The detailed grounds for this will be outlined in Chapters 2–6, where it will become apparent that many recent advances in quantum theory and relativity bear the stamps of being invented rather than discovered. Theoretical physics, in particular, now bears a close relationship to human practices usually described as arts, such as composing a symphony, creating a painting or writing a poem. Eddington was the first person of stature to propose the view that science is at least partly subjective, and it was put forth mainly in two volumes by him of a philosophical type (Eddington 1928, 1939). This view was met with something like respectful puzzlement by some physicists (Whittaker 1951, Dingle 1954). And it was met with outright hostility by several philosophers (Stebbing 1937, Nerlich 1967). However, a modern reading of the opinions of the latter shows that their criticism was mainly directed at how things were stated rather than the meaning of the statement; and today Eddington's views meet with more respect (Leslie 2001, Price and French 2004). If there is still a divide between science and the arts, it is narrower now than at any point in history.

Einstein, whose general theory of relativity was presented to the English-speaking world by Eddington, said that imagination is more important than knowledge (Figure 1.3). Everybody agrees that

**Figure 1.3.** Einstein, who spent his later years at Princeton, believed that the mind's powers of imagination are superior to its ability to store data.

imagination is an essential feature in the arts, and it is instructive to see how it figures in science.

That an act of the human mind is involved in science is evident even at the simple level of Newtonian mechanics. Let us reconsider the case of motion in a straight line (see above). Then Newton's laws tell us that the distance $s$ that a particle travels in time $t$ is give by $s = vt$, where $v$ is the velocity. This elementary relation already presumes that the natural state of an object is to continue in motion.

This may not have been obvious to the common person in the England of Newton's age, when a road was likely to be a muddy track in which a cart would come to rest unless encouraged to move by the force of horses or oxen. The state of continuous motion implied by the noted relation is more akin to that displayed by a ball rolling on a smooth table top. But even in the latter situation, friction brings the moving object eventually to rest. Thus the most basic law of motion we possess actually involves a somewhat counter-intuitive choice. We now admit that it is basically correct, given data on particles moving in vacuum tubes or satellites orbiting in space. However, the law involves a visualization of a state that is not common in everyday life, to which it is reconciled only by the invention of a countervailing force that we call friction. That is, Newton's laws involve an element of human insight which is close to what we call imagination.

The law $s = vt$ noted above can more instructively be written as $v \equiv s / t$, which defines the velocity. Here $s$ and $t$ are examples of what in basic physics are called extrinsic measures, while $v$ is an example of an intrinsic measure. Extrinsic measures are those whose values are divided when we divide the amount of the quantity under consideration. They include distance, time and mass. Intrinsic measures, by comparison, retain their values when we divide the amount of the quantity under consideration. Examples are density, pressure and temperature. The distinction between these classes is often overlooked in advanced physics, such as general relativity. But it is still present, because extrinsic measures are usually employed as the independent variables in a problem, while intrinsic measures are usually employed as the dependent variables. The distinction is present, for example, in Einstein's field equations for the behaviour of matter in a gravitational field. There the coordinates are frequently labelled $x$, $y$, $z$ and $t$ for space and time; while the properties of matter are commonly taken to be $\rho$, $p$, $T$ for density, pressure and temperature. The object of the exercise, in solving Einstein's equations, is to obtain the intrinsic measures as functions of the extrinsic ones. This is what we mean by a solution, say for the density of the galaxies $\rho = \rho(t)$ as a function of time since the big bang. At a basic level, the equations of physics are set up by making a choice between intrinsic and extrinsic measures, and this choice is essentially subjective.

Dimensional homogeneity is another property of the equations of physics which is often taken for granted but is basically subjective. We will discuss the meaning of dimensions in detail in Chapter 3. Here we note that it is universal in physics to categorize quantities in terms of the base dimensions, which for mechanics are denoted $M$, $L$, $T$ for mass, length and time. Thus the velocity $v \equiv s/t$ discussed above necessarily has the physical dimensions of $LT^{-1}$. Similarly, the density $\rho$ necessarily has the physical dimensions of $ML^{-3}$. Other quantities have more complicated dimensions. But the dimensional content of the terms in an equation of physics is always the same, meaning dimensional homogeneity. This property was at one time seen as puzzling, but is now recognized as an elementary application of group theory (Bridgman 1922, Wesson 1992). Also, since all the terms in an equation have the same physical dimensions, we can divide through by this and obtain an equivalent equation in which all the terms are dimensionless (i.e., they are pure numbers). Such quantities have the useful attribute of retaining their numerical size under changes of units, which are merely man-made standards for measuring things like mass, length and time. Also, the dimensionless quantities of real-world physics can be brought into correspondence with the numbers of abstract mathematics. This connection can in principle be used in reverse, and Eddington especially argued that much of physics might in principle be deducible from number theory. A less ambitious usage of the dimensionless quantities of physics is to reformulate the Cosmological Principle, so that it means not merely that the universe should "look the same" to all observers, but have physically-constructed dimensionless parameters which are measured to be the same by all observers (Wesson 1978). But however we use the dimensional homogeneity of the equations of science, it should be recalled that the assignment of physical dimensions to quantities is essentially subjective.

A critic might respond to the contents of the previous paragraphs by asking: "If you think that the equations of physics are subjective in nature, or at least partly the result of human imagination, then why do you trust them? Are not the equations of physics just a kind of distillation of common sense?"

This critic is mostly right in what he says, but probably wrong in why he says it. A person is justified in believing in the equations of physics, and these do mainly agree with common sense, but only as ideals that have to be qualified in application to real life. (For example, we can believe in Newton's laws of motion, but only when these are modified by the inclusion of friction.) It is wrong to believe that the laws of physics are sacrosanct. Certainly, they are not edicts of the kind found in the Bible. The average, practising scientist is not like the religious zealot who is dedicated to scripture. (There are a few scientific zealots, but their views are distrusted by the majority.) Rather, science has strength because its practioners are willing to take periodic looks at its foundations and ask if they are sound. And a good theoretician, in any field of science, must be willing to abandon a line of research if it proves invalid.

The amount of time and energy invested in producing a typical research paper is often underestimated by the non-scientist. The starting point of a new project is frequently an idea, which may be of a technical kind in an experimental area or of a more philosophical kind if the researcher is in a theoretical area. Today, most ideas are actually developed by more than one individual, and include graduate students, colleagues and sometimes technical personnel. Of the order of a hundred people may be involved in large projects, like mapping the human genome, searching for elementary particles or carrying out an astronomical survey. It is a non-trivial job to keep everybody 'in the loop' for months or years, and to coordinate their activities so that the research progresses in the most productive manner. Eventually, when the results are at hand, these are written up in a paper. The task of writing a paper is detested by many scientists. In large groups, the designated scribe is sometimes rewarded by first place in the list of authors. Otherwise, the general rule is that the authors' names appear alphabetically. A departure from this usually means that one researcher has made an exceptionally large contribution; but it may also indicate that a supervisor is pre-empting the work of graduate students or others with a lower place on the academic ladder. Such abuses happen, as do misuses of the refereeing process. The latter consists in sending the paper to one or more anonymous peers by the editor of the journal to which it is submitted. Though it is not common, there have been cases where the unknown referee has usurped the results in an article, while delaying an official response to it. The refereeing process is the most contentious part of the obstacle course through which the author has to steer a paper if it is to be published by a regular, hard-copy journal. Not surprisingly, some scientists prefer to short-cut the system, by sending the article to an electronic website. There, it can be read by all. However, this 'democratization' of science also brings with it many papers that are badly written, have poor logic or are just plain wrong. Many of the articles on websites will — after revision in accordance with readers' comments — be ultimately sent to regular journals. Assuming that the journal referees eventually recommend publication of the paper, the editor will send it to the printer. To avoid typesetting or software errors, the printer will usually send a preliminary copy or proof of the article to the first-listed author. When this is returned with corrections as needed, the paper is finally printed. It will be available to the general populace, either in a library or via an electronic version of the journal. Given the rigmarole of the publishing business, it is hardly surprising that from writing a draft to the appearance of the final version, a paper is typically delayed by six months to a year. In fast-moving areas of science such as genetics and cosmology, research results can be obsolescent before the public learns about them.

Despite the time and effort involved in publishing a scientific paper, more are being produced now than ever before. In the century from 1900 to 2000, research went from being the occasional occupation of the intellectual to being the staple of the modestly-educated person. Science has become an industry. And like other things which are mass-produced, the question arises of quantity versus quality. Although there is a wide variation, a productive researcher might be responsible for a couple of hundred papers during a career. However, a perusal of the journals today shows a host of articles which add an increment of insight to a hypothesis or a decimal point to a numerical result, but a dearth of papers which have a genuinely new idea or an original calculation. This is particularly the case in medicine and physics. In fact, the contents pages of most scientific journals have become so nit-pickingly technical as to be indecipherable to the average person. This may mark the demise of common sense as the basis for science. For why the outcome of some scientific calculation may be consistent with other knowledge or agree with experiment, it cannot be considered "common sense" in the true meaning of that phrase if it cannot be understood by the average or common man/woman. In a way, the credibility of science is threatened by its own cleverness.

Even professional scientists distrust things which appear to be too clever or abstract. There are many physicists who believe implicitly in

Newton's laws (and indeed trust their lives to them every day when driving home), but are uneasy about Einstein's laws. Yet the two sets are supposed to be connected by a secure line of reasoning. To compress the argument: Newton's laws of motion plus gravity need to be modified by the introduction of the invariable speed of light (special relativity), and the separate labels of space and time need to be joined into a manifold (spacetime) which is moreover curved by matter (general relativity), so that the force of gravity becomes the curvature of an imaginary surface. This statement is a condensation of what takes a couple of hundred pages to write out in detail. But whether in short form or long form, there are many people with a professional training in physics who will agree with the starting point but distrust the conclusion. The majority of these are not, by the way, 'cranks'. The latter are those who decline to listen to any argument which gainsays their own narrow viewpoint. By contrast, many reasonably open-minded folk find it difficult to follow the train of thought which, in effect, goes from a bouncing soccer ball to a singular black hole. For many people, common sense is lost somewhere along the way.

The concept of common sense is, in fact, a slippery one. Opinions about what is 'obvious' differ from person to person; and even if there is consensus about what is sensible at some point in history, it will more often than not change with time. In pre-Copernican days, it was apparently 'obvious' to most people that the Sun went around the Earth; but today an individual holding such an opinion would be called an idiot or a lunatic.

We should, however, be careful not to use the follies of history to give the impression that our ancestors were uniformly stupid. For example, it is frequently implied that scientists and philosophers of the past believed that the Earth was flat. This is incorrect. Our ancestors had the opportunity throughout prehistory to observe the phases of the Moon, which are the semicircular shapes produced when sunlight strikes a spherical body. Even though the physics may not have been clear to the average cave dweller, the fact of the circular shape must have been obvious. There are indications from archaeology that the original Indian inhabitants of North America could also see the phases of the planet Venus. This is not so surprising, when we recall that their eyes were more acute than those of the modern urbanite, who is more accustomed to seeing a street light than a planet. It is also reported that a man with sharp eyes, keeping watch from the top of a mast on a ship at sea, could detect the curvature of the horizon. Plato, in the pre-Christian era, wrote about the circle and the heavens. And of course the history of humankind is punctuated by observations of eclipses, when the Sun's disk is cut by the circular shape of the Moon, or when the circle of the Earth's shadow is cast onto its face. So, we realize on reflection that our forebears were not all card-carrying members of the flat-Earth society. In the modern Monty Python movie *The Meaning of Life*, the story of men's silly beliefs begins with a galleon that sails over the edge of the world into oblivion. But that is where the idea belongs: in fantasy. People in the past had their own versions of common sense, which while we may not endorse them today were nevertheless reasonable by ancient standards. Our ancestors were not morons. It is just that views of what is sensible have changed through time.

Do we really expect that the science of today will also be the accepted norm a hundred years hence? Almost certainly not. Assuming it does not defeat itself by trivial complexity (see above), science appears to have an open future. In this regard, it is like the arts, where there is always a new vogue in waiting. Indeed, science could probably only be halted by some significant sociological shift. This might be of the catastrophic variety, where society as a whole would be frozen by some natural or man-made calamity, maybe associated with global warming. Or it could be of the insidious variety, where society decides that new science is not desirable, such as might happen if experiments become too expensive or have potentially negative consequences. The Large Hadron Collider, which was completed in the fall of 2008, provides an example in the latter class (Figure 1.4). Its cost was around 10 billion dollars, which is comparable to the gross domestic product of a small country; and its high-energy collisions were feared by some, who argued that they could lead to the spawning of tiny black holes, which might eat up the Earth! However, while it is possible to imagine scenarios whereby the progress of science is halted on the experimental/observational front, it is unlikely that it can be stopped on the theoretical front. Indeed, many people think of "science" as shorthand for the spirit of inquiry and the urge to understand which separates humans from animals.

**Figure 1.4.** The large Hadron Collider is an expensive gadget which may be one of the last flings of experimental science.

Eddington was a quiet champion of the power of the human mind, a belief he shared with that of his contemporary Einstein. The latter is, of course, recognized as the paramount thinker, especially in regard to the foundation of the special and general theories of relativity. Later, we will examine these accounts in some detail. But for now, all we need to know is that the special theory describes events as affected by velocities; while the general theory extends to accelerations and forces, notably that of gravity (and by implication, masses). However, while the effects of relativity are now well understood, it is still a question of controversy as to whether Einstein discovered or invented it. Specifically, it is unclear whether or not Einstein was aware of the results of the Michelson–Morley experiment (see Chapter 7). This is commonly regarded as the breakthrough observation, which showed the invariance of the speed of light, and the non-existence of the medium (aether) which was supposed to support electromagnetic waves. The question of whether Einstein was aware of experiments that supported his theory of relativity is not only of interest to historians of science. For the larger question — of whether science is discovered or invented — goes to the heart of the subject, affecting both its contents and how we carry it out. On this question, Eddington (1928, 1939) wrote at length and with remarkable insight. He was of the opinion that science is largely invented.

The allegory of the fisherman and his net is one which is often quoted as illustrating Eddington's views. The fisherman has a net with a certain mesh dimension, and on retrieving his catch he notices that all of the fish have a minimum size, a rule he (wrongly) attributes to the sea and its contents, whereas it is actually a property of his net. Eddington applied this and other allegories to the sciences, arguing that they are at least partly subjective in nature. His philosophical views have sometimes been misinterpreted, and he certainly did not believe that the world is created inside our own heads, like the solipsist. But while he admitted the existence of an external world, he was convinced that our interpretation of it is necessarily conditioned by the biological and mental traits which attach to us being human. It is in this context that we should understand his much-quoted statement: "To put the conclusion crudely — the stuff of the world is mind-stuff."

In the following chapters, the aim is to inquire how far this provocative statement holds up in the context of modern science. There have assuredly been great changes in the mathematical sciences since Eddington's time, notably in quantum mechanics and cosmology (Bell 2004, Wesson 2007). It is now widely accepted that the physical sciences, at least in regard to how they are discussed, contain a cultural element (Shapin 2009). The biological sciences, also, have undergone a vast development (though Eddington was sympathetic to these, arguing that there is less interpretation intervening between the thing being observed and the person doing the observing). In the following five chapters we will concentrate on the 'hard' sciences. It is already clear that if Eddington's allegory of the fisherman's net is to be applied today, we will have to replace his single net by a suite of them — with the mesh sizes and shapes necessary to 'catch' the quantities of modern science. Our account will be quite concrete: we will, for example, ask just what is meant by things like the density and pressure of matter, which are used glibly by the physicist but whose origin we need to pin down. In this inquiry, we will perforce need to employ the occasional equation. But for the non-mathematically inclined, these should be regarded as shorthand for wordy statements, somewhat in the way a cartoon is used to convey the essence of a political argument. For more complicated equations — like Einstein's for general relativity — they can be regarded as paintings in a gallery, to be viewed and registered by the mind, before it moves on to consider other things. (Every equation is in any case accompanied by an explanation in words, as accurate as can be achieved by that medium.) Talking of works of art, we will frequently run across parallels between these and the products of science. We will also draw comparisons with music and literature, and briefly investigate that most thorny of subjects, the overlap (or lack of it) between science and religion. For science is an integral part of the culture of the modern world, and it is legitimate to ask how it relates to the more intuitive aspects of human thought.

In the present chapter, we have given an account of the warp of science. This means the basic laws and structure of it, as presently understood by the majority of scientists. In the following five chapters, we will sort through the material which is to be added to the warp, identifying the components of the scientific weft. This process is intricate and fascinating. The weaver who aims to produce a garment on a loom can set up the warp from any basic material, but the colour and texture of what he creates depends on picking through balls of wool or cotton for the weft, a process which involves choice. (Our scientific weft will be examined in the last chapter.) Likewise, the scientist who aims to give a complete theory of some natural phenomenon is faced throughout by issues of choice. In the case of a great scientist like Einstein, it is as if he set out single-handedly to weave the Bayeux Tapestry. We need to inquire how such things are achieved.

We particularly need to inquire about the issue to which Eddington drew attention: between the external world and the scientific account of it, there is a marvellous but poorly-understood filtering device, namely the human mind.

## References

Bell, J.S., 2004. Speakable and Unspeakable in Quantum Mechanics, 2nd edn. Cambridge University Press, Cambridge.

Bridgman, P.W., 1922. Dimensional Analysis. Yale University Press, New Haven.

Dingle, H., 1954. The Sources of Eddington's Philosophy. Cambridge University Press, Cambridge.

Eddington, A.S., 1928. The Nature of the Physical World. Cambridge University Press, Cambridge.

Eddington, A.S., 1939. The Philosophy of Physical Science. Cambridge University Press, Cambridge.

Halpern, P., 2000. The Pursuit of Destiny: A History of Prediction. Perseus, Cambridge, Mass.

Halpern, P., 2004. The Great Beyond: Higher Dimensions, Parallel Universes, and the Extraordinary Search for a Theory of Everything. Wiley, Hoboken, N.J.

Halpern, P., Wesson, P.S., 2006. Brave New Universe: Illuminating the Darkest Secrets of the Cosmos. J. Henry, Washington, D.C.

Hoyle, F., 1966. October the First is Too Late. Fawcett-Crest, Greenwich, Conn.

Hoyle, F., 1994. Home is Where the Wind Blows: Chapters from a Cosmologist's Life. University Science Books, Mill Valley, Cal.

Kramer, D. Stephani, H., MacCallum, M., Herlt, E., 1980. Exact Solutions of Einstein's Field Equations. Cambridge University Press, Cambridge.

Leslie, J., 2001. Infinite Minds: A Philosophical Cosmology. Clarendon, Oxford.

Nerlich, G.C., 1967. A.S. Eddington, *In* The Encyclopedia of Philosophy (ed. Edwards, P., vol. 2). Collier-Macmillan, New York, 458.

Price, K., French, S. (eds.), 2004. Arthur Stanley Eddington: Interdisciplinary Perspectives. Centre for Research in the Arts, Humanities and Social Sciences (10–11 March), Cambridge.

Shapin, S., 2009. Science as a Vocation. University Chicago Press, Chicago.

Stebbing, S., 1937. Philosophy and the Physicists. Methuen, London.

Wesson, P.S., 1978. Astron. Astrophys. 68, 131.

Wesson, P.S., 1992. Space Science Rev. 59, 365.

Wesson, P.S., 2007. Space-Time-Matter: Modern Higher-Dimensional Cosmology, 2nd edn. World Scientific, Singapore.

Whittaker, E.T., 1951. Eddington's Principles in the Philosophy of Science. Cambridge University Press, Cambridge.

Chapter 2

# PUZZLES OF PHYSICS

## 2.1 Introduction

To the conscientious physicist, a paradox is a poisonous thing. After all, if a theory is logically constructed, complete and in accordance with the known data, then its development and application should not present any contradictions.

So much for perfection. The average human cosmologist falls short of Laplace's imaginary super-being, who could comprehend all and predict everything (Laplace 1812). Instead we have the incubus of the apparent contradiction. But just as a physician can learn about health by studying disease, so can the physicist strengthen his position by examining and resolving paradoxes. It is with this positive attitude that we approach the major conundrums which face modern science.

## 2.2 Olbers' Paradox

This is the most notorious conundrum in science. If the universe is infinite and uniformly populated with luminous galaxies which have existed forever, then the night sky should be ablaze with light. Obviously it is not — but why?

The paradox actually predates Olbers, who however drew attention to it in 1823 (Figure 2.1). The argument for a bright night sky is simple and geometrical: in a uniform space, the volume goes up as the distance cubed, whereas the brightness of any source goes down as the inverse square, so the distant sources should predominate.

Possible resolutions of the paradox abound in the history of astronomy. Unfortunately, most of them are wrong. Olbers himself tended to the view that the intensity of light from distant regions of the universe was reduced by absorption due to matter in space. However, the conservation of energy tells us that even if there were significant amounts of such matter, the energy would be merely absorbed and re-radiated at other wavelengths, thereby shifting but not solving the problem.

In fact, two people of unlikely backgrounds did come close to the true solution. The first was a Swiss nobleman, Jean-Philippe Loys de Cheseaux in 1746, and the second was the American poet Edgar Allan Poe in 1848. Both realized that there was something amiss with the fundamental assumptions that underlie the paradox. The route to a resolution took a mistaken path, though, in 1952 when Bondi published an influential book on cosmology. Bondi was one of the originators of the steady-state theory, in which the universe always looks the same because matter is created in the void and comes together to form galaxies, thereby preserving the uniformity of the universe not only in space but also in time. This is certainly one of the most original and beautiful ideas in cosmology, and many astrophysicists still lament its demise from confrontation with data on

**Figure 2.1.** Olbers was a Prussian astronomer who puzzled about the darkness of the night sky and is thereby associated with the longest-running paradox in physics.

the microwave background, which is almost certainly the cooled-down radiation from the big bang fireball. (See Hoyle's autobiography of 1994 for a fascinating account of the genesis of the steady-state theory and other, more successful, developments in astrophysics in the latter half of the twentieth century. A related account is by Clayton 1975.) In Bondi's book, the darkness of the night sky was perforce attributed to the loss in energy of photons on their passage from their sources to the Milky Way due to the redshift effect. That a photon's energy is inversely proportional to its wavelength is an undisputable fact, so it was natural for Bondi in his considerations of a universe that was infinite in space and time to account for the darkness of intergalactic space by the redshift effect. Unfortunately, this explanation of Olbers' problem became fixed in the minds of many astronomers as the dominant and even unique one.

A little thought will show that this obsession is misguided. If the galaxies formed at some finite time in the past — as they would if the universe started in a big bang — then their stars would initially have been pumping photons into an intergalactic void that was black. The light from galaxies would start to fill space with radiation, even as the redshift effect acted in the opposite direction on its intensity. Also, the expansion of the universe meant that the volume of intergalactic space was increasing, so further diluting the energy density of the background field at optical wavelengths. That is, in the early universe two effects were competing: the emission from stars in galaxies was trying to brighten things up, while the redshift and expansion effects were trying to dim things down. These countervailing tendencies were realized by the British/American astrophysicist Harrison, who in the 1960s published a series of articles aimed at elucidating the problem, culminating in the appearance of a book which he hoped would set the record straight (Harrison 1987). However, Harrison's main line of reasoning was thermodynamical, and involved a balance of energies that was bolometric (summed over all wavelengths, so there was no specific figure available about the darkness of the night sky in the particular

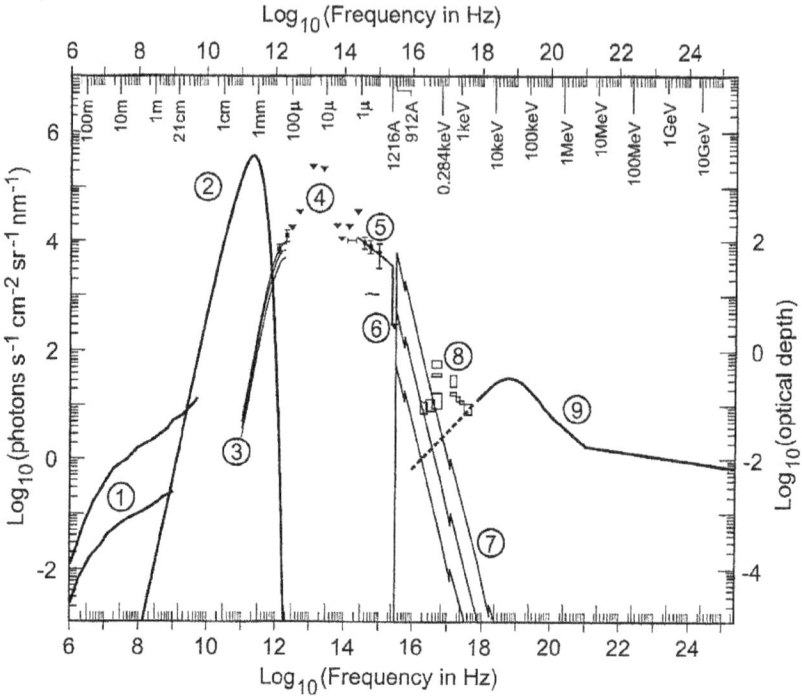

**Figure 2.2.** The spectrum of the electromagnetic radiation ('light') reaching us in various wavebands from all parts of the accessible universe. Shown are observational measurements and upper limits in the wavebands designated: (1) radio; (2) microwave; (3),(4) infrared; (5) optical; (6),(7) ultraviolet; (8) x-ray; (9) $\gamma$-ray. The microwave background (2) is believed to be the cooled-down radiation from the big bang; but the other backgrounds have astrophysical sources, including the extragalactic background light (5) that comes from stars in galaxies and puzzled Olbers.

waveband chosen by a given astronomer; see Figure 2.2). His arguments were therefore somewhat limited in their ability to persuade the majority towards a true understanding of the problem, even though he did appreciate that the effect of age was important.

Age can be seen to be important in two complementary ways. First, it directly limits the time over which galaxies have been pumping photons into intergalactic space, which influences the darkness of the night sky as seen from Earth. Second, if the galaxies have a finite age, the speed of light can be used to convert this to a distance, so in effect we only receive photons from within a certain portion of an unlimited universe, the size of that portion being determined by the age. (The distance/age relationship needs to be calculated carefully using relativity, but it turns out that the intensity of intergalactic radiation is surprisingly simple even in models based on Einstein's general theory.) Thus Olbers' problem is effectively one whose resolution involves various aspects of astrophysics, but most importantly age versus expansion.

Olbers' paradox was definitively resolved in 1987, using a new but straightforward method which separated the effects of age and expansion. (For a compressed account see the article by Wesson, Valle and Stabell 1987, and for a longer review see the book by Overduin and Wesson 2008.) The trick was to set up a realistic computer model of the light-emitting galaxies in an expanding universe, and then to stop the motion. This gave values for the intensity of intergalactic radiation with and without expansion, whose contending effects could thereby be evaluated. The results were clear and rapidly gained widespread acceptance: The darkness of the night sky is determined to order of magnitude by the age of the galaxies, and reduced by only a factor around 1/2 by the expansion of the universe.

## 2.3 Zero-Point Fields and the Cosmological 'Constant'

The concept of absolute temperature, with a zero point, was introduced by Kelvin. He realized that temperature was a measure of the energy of a system, and it is now widely accepted that all physical processes cease as the absolute temperature approaches zero. (Biological processes may stop at higher temperatures, especially if they involve water, which freezes at 273 degrees above absolute zero.) However, quantum mechanics as it is understood for higher-temperature systems, has a finite energy left for each element of a microscopic system when the temperature goes to zero. This applies even to fields which exist outside of ordinary matter or in vacuum. When the elements of a system are taken together, the result is an embarrassingly large energy density for these zero-point or vacuum fields.

Quantum field theory is based on the simple harmonic oscillator. A laboratory-sized example of such is the spring, which bounces on either side of a reference level, converting energy between the kinetic and potential forms. Imaginary oscillators of this type are taken as models for nearly all microscopic physical phenomena. For a field, the frequency $\omega$ defines the energy in conjunction with Planck's constant $\hbar$. (This is the straight value $h$ divided by $2\pi$, the factor reflects the traditional use of the angular frequency rather than the straight one defined as the inverse of the period.) The formal analysis gives the energy of the nth excitation or harmonic as

$$E_n = (n + 1/2)\hbar\omega. \qquad (2.1)$$

This agrees with observational data for $n > 0$. But the zero-point ($n = 0$) contribution of $\hbar\omega/2$ per frequency mode gives a large (and possibly infinite) energy when summed over modes. Correspondingly, when the three-dimensional size of the system is taken into account, there is an enormous energy density for the zero-point fields.

The problem is that such enormously energetic vacuum fields are not observed in nature. There appears to be a conflict between theory and observation of a troubling fundamental type.

The magnitude of the conflict depends on the kind of system concerned. Quantum field theory can be applied to any kind of system, and the problem has been studied in detail for electromagnetism and gravitation.

Electromagnetic zero-point fields are especially worrisome, because we pride ourselves on having an excellent classical theory following Maxwell and a very good quantum theory following Dirac. (Quantum electrodynamics, which deals with the interaction of electromagnetic fields and particles like the electron, is the best-verified segment of physical theory.) Yet if we take equation (2.1) above and sum over frequencies, we are led to the conclusion that the universe should posses an electromagnetic field more intense than those of the microwave background (due to the big bang) and the optical background (due to light from stars in galaxies). Indeed, the energy density of this field — assuming it gravitates in the same manner as ordinary photons in the manner described by Einstein's theory of general relativity — should cause a severe curvature in the spacetime of the universe. This is in conflict with observations of the gravitational lensing of objects like quasars. Even if we introduce a cutoff in the spectrum of the zero-point field given by (2.1), it would cause a break in the accurately black-body spectrum of the microwave background that is not observed (Wesson 1991). There are several ways out of this impasse. One, of course, is to say that the basic quantum field theory and its consequence (2.1) is just wrong. This is unpopular but conceivable, and will be examined below. However, another way to reconcile standard theory and observation is to assume that the electromagnetic zero-point field is real, but that its constituent photons behave in an anomalous manner, and do not gravitate. This idea is unorthodox, but its implications have been followed by researchers like Haisch, Puthoff and Rueda. They are motivated by an old suggestion due to the Soviet physicist Sakharov, who even argued that gravitation is due to a kind of 'shadowing' effect involving objects immersed in a zero-point field.

Gravitation appears to be the dominant interaction for the universe in the large. The best theory we have for this at present is general relativity, but it is classical in nature. Many researchers believe that it will break down at a quantum scale, given heuristically by combining the gravitational constant $G$, the speed of light $c$ and Planck's constant of action $\hbar$. This combination of parameters leads to a connected set of length, time and mass units. In ordinary measure, these have sizes of $1.6 \times 10^{-23}$ cm, $5.4 \times 10^{-44}$ s and

$$m_P = (\hbar c / G)^{1/2} \approx 2.2 \times 10^{-5} \text{g}. \qquad (2.2)$$

This is the Planck mass, which is widely regarded as roughly demarking the domains of classical and quantum gravity. (Though it should be said that some workers think this combination of parameters represents a naive approach to the quantization of gravity, a view supported by the fact that the universe is not dominated by $10^{-5}$ g black holes.) Assuming that a cutoff exists in the spectrum of the zero-point field at a wavelength given by the Planck length, it is straightforward to calculate the typical energy density of this kind of vacuum. It is of the order of $10^{112}$ erg cm$^{-3}$. This is gigantic by any standards. By contrast, the energy density of the universe in the large is currently thought to be set by the size of the cosmological constant. This itself has some paradoxical qualities, which we will discuss below. But taken at face value, the size of the cosmological constant from astrophysical data implies a corresponding energy density of order $10^{-8}$ erg cm$^{-3}$. The discrepancy, theory versus observation, is a mere $10^{120}$.

Numbers like these give even a cosmologist pause for thought. It is conceivable that the basic quantum theory that has to do with the simple harmonic oscillator is in error. Specifically, it is possible that the basic energy formula (2.1) should not contain the aberrant $\hbar\omega/2$ contribution from the zero-point field (or that it is in some way cancelled by another contribution). But while the reference level of energies can be reset for mechanical systems in the laboratory, the presence of a gravitational field in the large-scale universe makes this procedure awkward (Carroll 2004, pp. 173, 382). It is more likely that, while the component parts of our physical theory are valid in themselves, their combination involves subtleties of which we are presently unaware (Halpern 2004, Halpern and Wesson 2006). We will examine elsewhere the properties of unified field theories, particularly in regard to the influence of a mass-related scalar field that complements the effects of the gravitational and electromagnetic fields. Here, we admit that zero-point fields present a puzzle.

The solution to this puzzle will almost certainly involve a better understanding of the cosmological constant, as noted above. This parameter $\Lambda$ has many physical faces. As it appears in Einstein's theory, it is a true constant, on the same footing as the gravitational constant, the speed of light and Planck's constant of action or energy $(G, c, h)$. It has physical dimensions of an inverse length squared, and observations of supernovas and other objects show that the length involved is of the order of the size of the observable universe, namely $10^{28}$ cm. Alternatively, if we use the speed of light $c$ to convert from a length to a time, the characteristic number is of the same order as the age of the universe, namely $10^{10}$ yr. Also from Einstein's theory, the cosmological constant modulates a force per unit mass (or acceleration) that acts between any two objects in the universe. This is given in terms of the radial separation by $\Lambda rc^2 / 3$, and is repulsive for $\Lambda > 0$ as indicated by astrophysical data. (Though $\Lambda < 0$ is also allowed, and has been used by particle physicists to model small-scale phenomena such as quantum tunneling.) When Einstein's theory of the gravitational interaction is coupled to matter, it is also possible to interpret $\Lambda$ as an effective density and pressure for the vacuum, which we loosely take to mean the absence of ordinary matter. This density is $\Lambda c^2 / 8\pi G$, which on multiplication by $c^2$ gives the $10^{-8}$ erg cm$^{-3}$ quoted above for the energy density of the empty parts of the universe. These different ways of viewing the cosmological constant are discussed at greater length elsewhere. But it is clear that many branches of physics will need adjustment if it should turn out that $\Lambda$ is not a true constant.

This is a real possibility. The natural physical units of the cosmological 'constant' suggests that it might (to a first-order approximation) decay as the inverse square of the time elapsed since the big bang. Such a behaviour is compatible with astrophysical data; and more detailed models show that $\Lambda$ may have been formally infinite at the start of the universe and have diminished over our $13 \times 10^9$ yr history to the small value we observe today (Overduin 1999; Overduin, Wesson and Mashhoon 2007). Other models go further, and suggest that the cosmological 'constant' is in fact just the average value over large scales for a field that fluctuates in space. In such models, $\Lambda$ is replaced by a field of scalar type, which is strong around particles but weak over intergalactic distances. In this manner, the puzzle of zero-point fields and the problem of the cosmological 'constant' can both be solved, at least in principle.

## 2.4    The Hierarchy Problem

A question related to what we just discussed is why elementary particles are observed to have masses of order $10^{-24}$ g or less, when the value suggested by theory is the Planck mass of (2.2) or $m_P = (\hbar c / G)^{1/2} \simeq 2.2 \times 10^{-5}$ g. This question is related, via field theory, to the alternate one of why the interactions of particles are so much stronger than that due to gravity.

There are several potential answers to these questions, which together constitute the hierarchy problem. Pragmatically, they can be grouped into the complicated and the simple (though they are not mutually exclusive). Since we have already introduced many of the relevant concepts, we keep our considerations brief.

If the world has more than the four dimensions of spacetime — as seems increasingly likely — then conventional physics may operate on a surface in a manifold or 'space' of higher dimensions. This is currently the best option to unify the four known interactions, namely the strong, weak and electromagnetic forces of particles, plus gravity. (The subject of dimensions is discussed at length in Chapter 3.) The simplest extension of Einstein's theory of general relativity is to five dimensions, and this has been studied in the versions known as membrane theory and space-time-matter (or induced-matter) theory. In membrane theory, 4D spacetime is a singular surface in the 5D manifold. Particles are crammed into this thin layer with consequently strong interactions, whereas gravity can operate outside it in the 'bulk' and is so weaker (Randall 2002). The philosophy here is to divorce gravity from the interactions of particles, leaving the masses to be determined by local physics.

A simpler approach, and more global, is to treat the fifth dimension on the same basis as the other four, and identify the extra one as a geometrical description of mass (Wesson 2008). For ordinary 3D space, we can describe distances in different ways: for example, by Cartesian coordinates $xyz$ or spherical polar ones $r\theta\phi$. Both will give the same answer if the theory is set up using tensors, which are invariant under a change of coordinates (see elsewhere). For the fifth coordinate, nature provides us with two ways to measure the mass of an object in terms of a length:

$$l = Gm/c^2 \quad \text{or} \quad l = h/mc. \tag{2.3}$$

These are the Schwarzschild radius and the Compton wavelength. The former is gravitational while the latter is quantum mechanical. And just as it makes no sense to mix the ways of measuring a 3D distance ($xyz$ versus $r\theta\phi$), it makes no sense to mix the two ways of measuring mass. If we do, then we obtain the Planck mass (2.2). However, a physicist should not expect to observe an object with a mass of order $10^{-5}$g any more than a fruit grower should expect to pick something that is half an apple and half an orange. According to space-time-matter theory, the mass of a particle is determined by the scalar field which forms the fifth dimension, in conjunction with a length set by one *or* the other of (2.3), depending on whether we use a classical or quantum unit of measurement.

## 2.5    Supersymmetry and Dark WIMPs

Symmetries are powerful ways to categorize the properties of particles, and point the way to a group of weakly-interacting massive particles (WIMPs) which may comprise as much as a quarter of the matter in the universe. Supersymmetry is a particularly broad concept, but a puzzling one, in that its strong theoretical base is not matched by observations.

In supersymmetry, the familiar particles of the Standard Model are matched by another family of massive objects whose spin properties are such as to cause a cancellation of the unacceptably large vacuum fields which exist otherwise. (See above: this is in effect a local solution to the problem of zero-point fields.) An example is provided by the spin-2 graviton, which in particle physics mediates the classical gravitational field as described by general relativity. With supersymmetry, this graviton is partnered with a hypothetical spin-3/2 particle called the gravitino. The application of this scheme to all of the known particle species causes an increase by at least a factor two in the number of 'elementary' particles. However, some of the new particles have interesting properties, most notably the WIMPs, which should hardly interact at all with other material but have masses large enough to account for the exotic dark matter inferred from observations of galaxies. The fact that there is scant evidence of supersymmetry in the present low-temperature universe is commonly dealt with by assuming that it was indeed present in the early high-temperature universe, but that the symmetry was broken by the cooling inherent to the expansion that followed the big bang. As the temperature of the fireball dropped below their rest energies, heavy species would have dropped out of equilibrium and begun to disappear through the process of pair annihilation, leaving progressively lighter ones behind. Eventually, only one massive superpartner would have remained. It is this lightest supersymmetric WIMP which is believed to make up most of the dark matter.

The preceding argument may sound plausible to a particle physicist, but is less so to a cosmologist versed in classical field theory. The main reason is that exotic dark matter — whether WIMPs or something else — represents only one of (at least) two unseen constituents of the universe. The other one is dark energy. This appears to make up about 74% of the stuff of the cosmos, based on its acceleration as revealed by supernova observations (Perlmutter et al. 1999, Astier et al. 2006). To a good approximation, the dark energy has properties similar to the cosmological constant of general relativity when that parameter is interpreted as a kind of fluid (see above and elsewhere for more detailed discussions of this topic). However, the cosmological 'constant' is actually a particularly simple example of a scalar field, which does not depend on direction or orientation as do the vector interaction of electromagnetism and the tensor interaction of gravitation. To the cosmologist conversant with general relativity, it seems natural to put dark matter and dark energy together, and explain the duo in terms of an extended version of that theory. A popular way to extend Einstein's theory is to add dimensions, as we have seen before. And it is only necessary to add one, to incorporate a scalar field which can account for both dark matter and dark energy (Wesson 2008). In contrast, the concept of supersymmetry as favoured by particle physicists requires the addition of many more dimensions. The most conservative such approach involves 10D. From the viewpoint of standard cosmology as based on 4D general relativity, this is because a 10D imaginary space that is flat is the simplest way to rewrite a 4D space that is curved by energy, thereby resolving the problem of vacuum fields. However, there are also valid arguments for considering spaces of even higher dimensions. Thus 11D splits naturally into 7D plus 4D, where the latter may be identified with spacetime. While 26D and other manifolds have algebraic properties that allow a point particle to be replaced by a string or other structure, where problems to do with divergent energies may be better resolved. A philosopher wielding Ockham's razor would find much to attack in these higher-dimensional approaches to physics, and it remains to be seen if supersymmetry will survive the barber.

## 2.6   The Fermi–Hart Paradox: Where are the Aliens?

Fermi is reported to have mused over lunch that there could not be intelligent lifeforms elsewhere than Earth because they would have colonized space and already be here. Conversely, the presence of life on the Earth implies its presence elsewhere. This problem, though it originated with Fermi, has been worked on by many people, most notably Hart. Over the years, the absence of evidence for aliens has emerged as one of the few solid data in the field of exobiology. In this section, we therefore confine our attention to the Fermi–Hart paradox and how it may be resolved.

An immediate suggestion for a resolution is, of course, that alien civilizations are sparse in the universe. Indeed, Tipler and others have argued that human civilization may be unique and that we are truly alone. This view is distasteful to many, and some researchers such as Clarke and Sagan have reached the opposite conclusion, that life (and by implication civilization) is common. The reason for this divergence of opinion lies simply in a lack of data.

Drake's formula is the traditional way to quantify the frequency of extraterrestrial civilizations. It involves a product of probabilities, ranging from the astrophysical (e.g. the fraction of stars which have habitable planets) to the sociological (e.g. the relative timescale for the development of technology). However, each of the component probabilities is poorly known, so the result has a high degree of uncertainty. The nearest technological civilization to us may be around a nearby star, or in one of the most remote galaxies.

It is important to realize in the application of Drake's formula that there is a continuum of systems to which it can be applied: the stars of the Milky Way, the local group of galaxies, or all of the objects in the visible universe. To order of magnitude, there are as many galaxies in that part of the universe accessible to observation as there are stars in the Milky Way. Signals from extraterrestrial civilizations involve, for their detection, a balance between the number of sources (which goes up approximately as the cube of the distance) and the signal strength (which goes down for electromagnetic radiation as the square of the distance). This is like the situation we encountered before in regard to Olbers' paradox. The American program on the Search for Extraterrestrial Intelligence has concentrated on nearby stars, but previous Russian surveys focussed on distant galaxies. The fact that no signals have been detected from either class of objects brings us back to the Fermi–Hart paradox.

It actually makes little difference to the veracity of this paradox whether we consider the detection of signals from extraterrestrial civilizations or other evidence of their existence. It has been estimated, for example, that a technologically advanced race could populate the Milky Way with robotic devices in a period of about $3 \times 10^6$ yr, which is small compared to the Galactic age of approximately $13 \times 10^9$ yr. The fact that we have found no evidence of aliens in our solar system reinforces the fact that we have received no signals from them. (This is the case even though planets orbiting other stars have recently been detected: see Marcy et al. 2005.) A few researchers, such as the radio astronomer Verschuur, have suggested that the money spent on S.E.T.I. might be better spent on alleviating problems on our own troubled planet. While most scientists would probably not go this far, it is becoming more imperative to face up to the question: Are we alone?

An answer to this, and a resolution of the Fermi–Hart paradox, is provided by cosmology. According to Einstein's theory of general relativity, the universe is isotropic and homogeneous ($\equiv$ uniform), with no centre and no edge. But it began in something like a big bang, approximately $13 \times 10^9$ yr ago. As with Olbers' problem, the fact it has a finite age, and that the speed of light has a finite value, means that we cannot see all of the universe at any given time. Our view is restricted by a kind of imaginary surface. To appreciate that such must exist, consider that as we view ever more distant galaxies, we observe them as they were at earlier times. If we could image them, we would eventually see the galaxies as they were at formation. And because our observations must be similar in all directions, that galaxy-formation place must be at the same distance from us in all directions in space. That is, it must form a shell around us (though this does not imply a centre, because our location is arbitrary). It we could peer further, we would merely see the amorphous medium from which the galaxies presumably condensed. And if we could look even further back, to a time about $13 \times 10^9$ yr ago, we could see the fireball which followed the big bang. In this picture, the big bang is smeared over a fiery shell around us; and because nothing existed before, that shell is a kind of ultimate surface. It separates what we can see from what we cannot see, and by analogy with the situation on Earth is termed the horizon (see Figure 2.3). The cosmological horizon, by its nature, is also a limit for the transmission of information by means of light. The precise distance to the horizon depends on the detailed properties of the universe, and given some simplifying assumptions can be written down (see Weinberg 1972 p. 489; Halpern and Wesson 2006). For the simplest case, it is $3ct_0$ where $t_0$ is the age ($\approx 13 \times 10^9$ yr). The fact this is not just the speed of light times the age is due to the influence of general relativity; and we remark in passing that many of the so-called paradoxes of special relativity arise in situations where accelerations or gravity are involved, for which the more complete theory is required for an accurate analysis. Here, we note that the distance to the horizon for the present universe is enormous by conventional standards, but still finite. More importantly, the data which go into Drake's formula, considered above, show that the nearest extraterrestrial civilization may be close to or even beyond the cosmological horizon.

In other words, aliens may exist but cannot communicate with us.

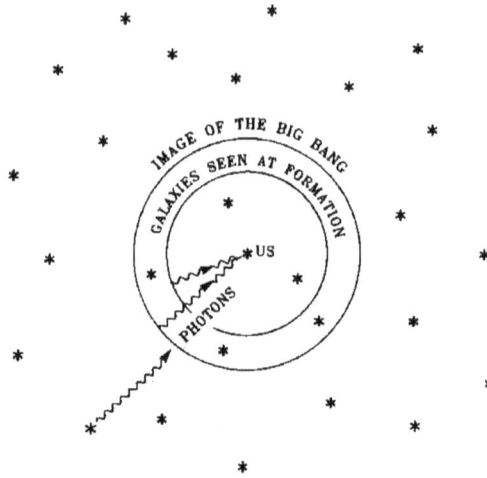

**Figure 2.3.** As we look further out into space, the time-lag associated with the finite speed of light means that we see back to events at earlier times. This implies that past events appear to us to be on the surfaces of imaginary shells drawn about us as (arbitrary) centre. Thus we can in principle receive photons from a surface where we would see galaxies at formation, and ultimately see an image of the big bang. The imaginary surface connected with the latter event is known technically as the (particle) horizon. The universe 'exists' beyond the horizon, but in analogy with its Earthly counterpart, we cannot see that far. That is, photons from beyond the horizon have not had time to reach us yet. This geometry comes ultimately from relativity, and in effectively transforming the explosion of the big bang into a fiery spherical shell, is of both physical and philosophical importance. If Olbers had understood this situation, he could have avoided the historical paradox of the dark night sky. And it may provide an explanation for the status of the Search for Extraterrestrial Intelligence, insofar as aliens may exist but lie beyond our cosmological horizon, so their messages may not have reached us yet.

This may seem to be a kind of political resolution of the Fermi–Hart paradox. And depressing to some. For the latter, and to be objective, let us briefly consider how the preceding conclusion might be avoided.

Two exits from the Fermi–Hart paradox are provided by the currently popular idea of extending general relativity from four to five or more dimensions, as a means of unifying gravity with the other interactions of physics. The extra dimension of 5D relativity is poorly understood, but is believed to be connected to a scalar field which augments the actions of electromagnetism and gravity (Wesson 2008). The extra coordinate is often assumed to behave like one of the directions of ordinary space, but it could also behave like a second axis of time. In this case, the speed in ordinary 3D space is not limited to that of light, but rather to $(1 + w^2 / c^2)^{1/2} c$, where $w$ is the 'velocity' in the extra dimension. This means that speeds in ordinary space can be superluminal, so extraterrestrials with a good understanding of 5D physics might be able to signal at speeds greater than previously assumed. A second, more subtle property of 5D relativity involves the concept of simultaneity. In 4D, this means in practical terms that particles exchange information by the passage of photons, which necessarily have zero rest mass. In 5D, the concept becomes broader, so that particles can be in causal contact even if they are massive. This raises the intriguing possibility that objects in the universe are already 'aware' of each other in a way which involves the fifth dimension, and makes obsolescent the conventional idea of an horizon. If this is so, S.E.T.I. may already be a done deal; but we will not be aware of the extra-dimensional chatter until we develop the appropriate technology.

There are, of course, other ways of explaining the apparent absence of extraterrestrials and their signals. Prime among these is the cosmic zoo hypothesis. According to this, the aliens are aware of us by virtue of our radio and television broadcasts, which now fill a sphere many light years in size that is large enough to encompass numerous stars. However, the aliens disdain contact with us; which given the nature of the average television program is hardly surprising.

In the above account, we have concentrated on those aspects of exobiology which are presently amenable to scientific study. The longer we go without evidence for aliens, the more pressing becomes explanation for that fact. However, there is a different aspect of the problem which is becoming increasingly profitable to study, and that is panspermia. This is the idea that life may be seeded among the stars by astrophysical means. It is due primarily to Arrhenius and has been around for over a century, but recent discoveries have given it a new impetus (Arrhenius 1908; Secker, Lepock and Wesson 1994). We can consider, for example, the ejection of dust grains with micro-organisms from one star system and their passage to another such. Do the organisms survive the vicissitudes of their journey so that they can seed life at their destinations? The answer is that a few hardy ones might do so. However, further thought shows that the problem has less to do with the organism itself than with the genetic information it carries. Life is, basically, a reproducing form of biologically-supported information. Dead organisms, provided they find an hospitable environment, can lead to the development of new life. It may be that a profitable subject for future study will be necropanspermia.

## 2.7    Conclusion

It is in the nature of scientific research that it throws up puzzles in its development. But when these grow into paradoxes, it is a sign that something is amiss in the process. To paraphrase Shakespeare: *the paradox is in the brain of the beholder, not the external world.*

This is plain from our study of Olbers' paradox (Section 2.2). This survived as a major conundrum for over a century because the underlying problem was poorly formulated; and because a possible solution (the expansion of the universe) became endemic to the neglect of the proper solution (the finite age of the universe). The fact that the poet Edgar Allan Poe came within a hair's breadth of resolving the issue shows that what was required was merely a modicum of clear, unbiased thought. A dose of the latter is needed also to resolve the puzzle of zero-point fields and the cosmological constant (Section 2.3). These issues are related to the hierarchy problem (Section 2.4). All revolve around a proper understanding of the relationship between macroscopic and microscopic physics. These may require different intrinsic scales for their description, rather than the mixed one that leads to the so-called Planck mass, which in the real world is conspicuous by its absence. Supersymmetry as a concept is closely related to whether the dark matter in the universe consists predominantly of weakly-interacting massive particles (Section 2.5). However, the dark matter could be more closely connected to dark energy, with both originating from a scalar interaction that for its proper description requires an extension of general relativity to five (or more) dimensions. Supersymmetry may turn out to be like the aether: believed in, never found and ultimately abandoned. Something else which is not found is evidence for aliens (Section 2.6). If such existed, they would arguably have colonized the Milky Way and be here; ergo they do not exist, even though the presence of life on Earth seems to imply its existence elsewhere. This paradox, named after Fermi and Hart, is similar in nature to that of Olbers, and may have a similar resolution. Extraterrestrial civilizations may exist in principle, but be located beyond the cosmological horizon associated with the finite age of the universe, and so be unable to communicate with us in practice. Alternatively — and much more simply — the aliens despise the television programs we leak into space, and have decided to quarantine us in a kind of cosmic zoo.

Irrespective of whether something is called a puzzle or a paradox, all such are eventually solved. Science is, after all, a logical activity. So while its practitioners may have occasional trouble in its practice, science must inherently be free of contradictions. This no matter how frustrating they may be. Consider, for example, the question "What happens when an irresistible force meets an immoveable object?" This may confound a child; but the adult will point out that once an irresistible force has been postulated, there cannot by definition be such a thing as an immoveable object. Many of the apparent paradoxes posited by Russell and other philosophers are at base of this type: carefully-worded entrapments. Of course, it is possible to identify more weighty problems, particularly ones involving modern physics. These include the question of how to quantize gravity (maybe it cannot be done, and quantum mechanics instead needs to be remodeled in the guise of general relativity); the issue of whether there was really a big bang (maybe it instead has the nature of a coordinate singularity that can be removed in a higher-dimensional theory of gravity); and the perennial problem of the origin of life (maybe the Earth was seeded by dead, information-carrying bits of biological material). These and other fundamental issues are fascinating to the researcher, who sometimes pursues them with a passion that is baffling to the non-scientist. It is the existence of such problems which lies behind Rutherford's oft-quoted and rather provocative remark, to the effect that science is physics and the rest is just stamp collecting.

## References

Arrhenius, S., 1908. Worlds in the Making. Harper and Row, London.

Astier, P., et al., 2006. Astron. Astrophys. 447, 31.

Carrol, S.M., 2004. Spacetime and Geometry: An Introduction to General Relativity. Addison-Wesley, San Francisco.

Clayton, D.D., 1975. The Dark Night Sky: A Personal Adventure in Cosmology. Demeter-Quadrangle, New York.

Halpern, P., Wesson, P., 2006. Brave New Universe: Illuminating the Darkest Secrets of the Cosmos. J. Henry, Washington, D.C.

Halpern, P., 2004. The Great Beyond: Higher Dimensions, Parallel Universes and the Extraordinary Search for a Theory of Everything. Wiley, Hoboken, N.J.

Harrison, E.R., 1987. Darkness at Night. Harvard University Press, Cambridge, Mass.

Hart, M.H., Zuckerman, B. (eds.), 1982. Extraterrestrials — Where are They? Pergamon, New York.

Hoyle, F., 1994. Home is Where the Wind Blows: Chapters from a Cosmologist's Life. University Science Books. Mill Valley, Cal.

Laplace, P.S., 1812. Analytical Theory of Probability. Courcier, Paris.

Marcy, G., et al., 2005. Prog. Theor. Phys. Suppl. 158, 1.

Overduin, J.M., 1999. Astrophys. J. 517, L1.

Overduin, J.M., Wesson, P.S., Mashhoon, B., 2007. Astron. Astrophys. 473, 727.

Overduin, J.M., Wesson, P.S., 2008. The Light/Dark Universe. World Scientific, Singapore.

Perlmutter, S., et al., 1999. Astrophys. J. 517, 565.

Randall, L., 2002. Science 296 (5572), 1422.

Secker, J., Lepock, J., Wesson, P.S., 1994. Astrophys Sp. Sci. 219, 1. [See also Wesson, P.S., 1990, Quart. J. Roy. Astr. Soc. 31, 161.]

Weinberg, S., 1972. Gravitation and Cosmology. Wiley, New York.

Wesson, P.S., Valle, K., Stabell, R., 1987. Astrophys. J. 317, 601. [See also Wesson, P.S., 1991. Astrophys. J. 367, 399.]

Wesson, P.S., 1991. Astrophys. J. 378, 466. [See also Wesson, P.S., 2000. Zero-Point Fields, Gravitation and New Physics. http://www.calphysics.org.]

Wesson, P.S., 2008. Gen. Rel. Grav. 40, 1353.

## Chapter 3

# THE MEANING OF DIMENSIONS

## 3.1    Introduction

Dimensions are both primitive concepts that provide a framework for mechanics and sophisticated devices that can be used to construct unified field theories. Thus the ordinary space of our perceptions (*xyz*) and the subjective notion of time (*t*) provide the labels with which to describe Newtonian mechanics. And with the introduction of the speed of light to form a time-related *coordinate* (*ct*), it is straightforward to describe Einsteinian mechanics. Used in the abstract, dimensions also provide a means of extending general relativity in accordance with certain physical principles, like 10D supersymmetry. As part of the endeavour to unify gravity with the interactions of particle physics, there has recently been an explosion of interest in manifolds with higher dimensions. Much of this work is algebraic in nature. Therefore, to provide some balance and direction, we will concentrate here on fundamentals and attempt to come to an understanding of the meaning of dimensions.

Our main conclusion, based on 40 years of consideration, will be that dimensions are basically inventions, which have to be chosen with skill if they are to be profitable in application to physics. This view may seem strange to some workers, but is not new. It is implicit in the extensive writings on philosophy and physics by the great astronomer Eddington, and has been made explicit by his followers, including the writer. This view is conformable, it should be noted, with algebraic proofs and other mathematical results on many-dimensional manifolds, such as those of the classical geometer Campbell, whose embedding theorem has been recently rediscovered and applied by several workers to modern unified-field theory. Indeed, a proper understanding of the meaning of dimensions involves both history and modern physics.

There is a large literature on dimensions; but it would be inappropriate to go into details here, and we instead list some key works. The main philosophical/physical ones are those by Barrow (1981), Barrow and Tipler (1986), Eddington (1935, 1939), Halpern (2004), Kilmister (1994), McCrea and Rees (1983), Petley (1985), Price and French (2004) and Wesson (1978, 1992). The main algebraic/mathematical works are those by Campbell (1926), Green et al. (1987), Gubser and Lykken (2004), Seahra and Wesson (2003), Szabo (2004), Wesson (2006, 2007) and West (1986). These contain extensive bibliographies, and we will quote freely from them in what follows.

The plan of this chapter is as follows. Section 3.2 outlines the view that dimensions are inventions whose application to physics involves a well-judged use of the fundamental constants. This rests on work by Eddington, Campbell and others; so in Sections 3.3 and 3.4 we give accounts of the main philosophical and algebraic results (respectively) due to these men, in a modern context. Section 3.5 is a summary, where we restate our view that the utility of dimensions in physics owes at least as much to skill as to symbolism. We aim to be pedagogical rather than pedantic, and hope that the reader will take our comments in the spirit of learning rather than lecture.

## 3.2    Dimensions and Fundamental Constants

Minkowski made a penetrating contribution to special relativity and our view of mechanics when by the simple identification of $x^4 \equiv ct$ he put time on the same footing as the coordinates $x^{123} = xyz$ of the ordinary space of our perceptions. (We will examine this in detail in Chapter 4.) Einstein took an even more important step when he made the Principle of Covariance one of the pillars of general relativity, showing that the 4 coordinates traditionally used in mechanics can be altered and even mixed, producing an account of physical phenomena which is independent of the labels by which we choose to describe them. These issues are nowadays taken for granted; but a little reflection shows that insofar as the coordinates are the labels of the dimensions, the latter are themselves flexible in nature.

Einstein was in his later years preoccupied with the manner in which we describe matter. His original formulation of general relativity involved a match between a purely geometrical object we now call the Einstein tensor ($G_{\alpha\beta}$, $\alpha$ and $\beta = 0,123$ for $t, xyz$); and an object which depends on the properties of matter, the energy-momentum or stress-energy tensor ($T_{\alpha\beta}$, which contains quantities like the ordinary density $\rho$ and pressure $p$ of matter). The coefficient necessary to turn this correspondence into an equation is (in suitable units) $8\pi G / c^4$, where $G$ is the gravitational constant. Hence Einstein's field equations

$$G_{\alpha\beta} = (8\pi G / c^4)T_{\alpha\beta} \ (\alpha, \beta = 0,123), \quad (3.1)$$

which are an excellent description of gravitating matter. In writing these equations, it is common to read them from left to right, so that the geometry of 4D spacetime is governed by the matter it contains. However, this split is artificial. Einstein himself realized this, and sought (unsuccessfully) for some way to turn the "base wood" of $T_{\alpha\beta}$ into the "marble" of $G_{\alpha\beta}$. His aim, simply put, was to geometrize all of mechanical physics — the matter as well as the fields.

A potential way to geometrize the physics of gravity and electromagnetism was suggested in 1920 by Kaluza, who added a fifth dimension to Einstein's general relativity. Kaluza showed that the apparently empty 5D field equations

$$R_{AB} = 0 \ (A, B = 0,123,4) \quad (3.2)$$

in terms of the Ricci tensor, contain Einstein's equations for gravity and Maxwell's equations for electromagnetism. Einstein, after some thought, endorsed this step. However, in the 1920s quantum mechanics was gaining a foothold in theoretical physics, and in the 1930s there was a vast expansion of interest in this area, at the expense of general relativity. This explains why there was such a high degree of attention to the proposal of Klein, who in 1926 suggested that the fifth dimension of Kaluza ought to have a closed topology (i.e., a circle), in order to explain the fundamental quantum of electric charge (e). Klein's argument actually related this quantity to the momentum in the extra dimension, but in so doing introduced the fundamental unit of action ($h$) which is now known as Planck's constant. However, despite the appeal of Klein's idea, it was destined for failure. There are several technical reasons for this, but it is sufficient to note here that the crude 5D gravity/quantum theory of Kaluza/Klein implied a basic role for the mass quantum $(\hbar c / G)^{1/2}$. This is of order $10^{-5}$ g, and does not play a dominant role in the spectrum of masses observed in the real universe. (See Chapter 2; whether we use $h$ or $\hbar \equiv h / 2\pi$ is not of fundamental importance, the choice being related to whether the application involves a simple or angular frequency.) In more modern terms, the so-called hierarchy problem is centred on the fact that observed particle masses are far less than the Planck mass, or any other mass derivable from a tower of states where this is the basic unit. In addition to this shortcoming, the extra dimension of Klein was supposed to be rolled up to a size that was unobservably small ('compactificaion'). We see in retrospect that the Klein modification of the Kaluza scheme was a dead end. This does not, though, imply that there is anything wrong with the basic proposition, which follows from the work of Einstein and Kaluza, that matter can be geometrized with the aid of the fundamental constants. As a simple example, an astrophysicist presented with a problem involving a gravitationally-dominated cloud of density $\rho$ will automatically note that the free-fall or dynamical timescale is the inverse square root of $G\rho$. This tells him immediately about the expected evolution of the cloud. Alternatively, instead of taking the density as the relevant physical quantity, we can form the length $(c^2/G\rho)^{1/2}$ and obtain an equivalent description of the physics in terms of a geometrical quantity.

The above simple outline, of how physical quantities can be combined with the fundamental constants to form geometrical quantities such as lengths, can be much developed and put on a systematic basis (Wesson 2007). The result is induced-matter theory, or as some workers prefer to call it, space-time-matter theory. The philosophical basis of the theory is to realize Einstein's dream of unifying geometry and matter (see above). The mathematical basis of it is Campbell's theorem, which ensures an embedding of 4D general relativity with sources in a 5D theory whose field equations are *apparently* empty (see below). That is, the 4D Einstein equations of (3.1) are embedded perfectly in the 5D Ricci-flat equations of (3.2). The point, in simple terms, is that we use the fifth dimension to model matter.

An alternative version of 5D gravity, which is mathematically similar, is membrane theory. In this, gravity propagates freely in 5D, into the 'bulk'; but the interactions of particles are confined to a hypersurface or the 'brane'. It has been shown by Ponce de Leon and others that both the field equations and the dynamical equations are effectively the same in both theories. The only difference is that whereas induced-matter theory treats all five dimensions as equivalent, membrane theory makes spacetime a special (singular) hypersurface. For induced-matter theory, particles can wander away from the hypersurface at a slow rate governed by the cosmological constant; whereas for membrane theory, particles are confined to the hypersurface by an exponential force governed by the cosmological constant (see Chapters 2 and 5). Both versions of 5D general relativity are in agreement with observations. The choice between them is largely philosophical: Are we living in a universe where the fifth dimension is 'open', or are we living an existence where we are 'stuck' to a particular slice of the 5D manifold?

Certainly, the fundamental constants available to us at the present stage in the development of physics allow us to geometrize matter in terms of one extra dimension. Insofar as mechanics involves the basic physical quantities of mass, length and time, it is apparent that any code for the geometrization of mass will serve the purpose of extending 4D spacetime to a 5D space-time-mass manifold. The theory is covariant. However, not all parametizations are equally *convenient*, in regard to returning known 4D physics from a 5D definition of 'distance' or metric. Thus, the 'canonical' metric has attracted much attention. In it, the line element is augmented by a flat extra dimension, while its 4D part is multiplied by a quadratic factor (the corresponding metric in membrane theory involves an exponential factor, as noted above). For the canonical metric, the physics flows from the factor $(l / L)^2$ where $x^4 = l$ and $L$ is a constant. The last can be evaluated by comparison with the 4D Einstein metric, giving $L = (3 / \Lambda)^{1/2}$ where $\Lambda$ is the cosmological constant. In this way, we weld ordinary mechanics to cosmology, with the identification $x^4 = l = Gm / c^2$ where $m$ is the rest mass of a macroscopic object. If, on the other hand, we wish to study microscopic phenomena, the simple coordinate transformation $l \rightarrow L^2 / l$ gives us a quantum (as opposed to classical) description of rest mass via $x^4 = h / mc$. In other words, the large and small scales are accommodated by choices of coordinates which utilize the available fundamental constants, labelling the mass either by the Schwarzchild radius or by the Compton wavelength.

It is not difficult to see how to extend the above approach to higher dimensions. However, skill is needed here. For example, electric charge can either be incorporated into 5D, along the lines originally proposed by Kaluza and Klein; or treated as a sixth dimension, with coordinate $x_q \equiv (G / c^4)^{1/2} q$ where $q$ is the charge, as studied by Fukui and others. A possible resolution of technical problems like this is to 'fill up' the parameter space of the lowest-dimensional realistic model (in this case 5D), before moving to a higher dimension. As regards other kinds of 'charges' associated with particle physics, they should be geometrized and then treated as coordinates in the matching $N$-dimensional manifold. In this regard, as we have emphasized, there are choices to be made about how best to put the physics into correspondence with the algebra. For example, in supersymmetry, every integral-spin boson is matched with a half-integral-spin fermion, in order to cancel off the enormous vacuum or zero-point fields which would otherwise occur (Section 2.3). Now, it is a theorem that any curved energy-full solution of the 4D Einstein field equations can be embedded in a flat and energy-free 10D manifold. (This is basically a result of counting the degrees of freedom in the relevant sets of equations.) It is the simplest motivation known to the writer for supersymmetry. However, it is possible in certain cases that the condition of zero energy can be accomplished in a space of less than 10 dimensions, given a skillful choice of parameters.

Physicists have chosen geometry as the currently best way to deal with macroscopic and microscopic mechanics; and while there are theorems which deal with the question of how to embed the 4D world of our senses in higher-dimensional manifolds, the choice of the latter requires intuition and skill.

## 3.3    Eddington and His Legacy

In studying dimensions and fundamental constants over several decades, the writer has come to realize that much modern work on these topics has its roots in the views of Arthur Stanley Eddington (1882–1944; for a recent interdisciplinary review of his contributions to physics and philosophy, see the conference notes edited by Price and French 2004). He was primarily an astronomer, but with a gift for the pithy quote. For example: "We are bits of stellar matter that got cold by accident, bits of a star gone wrong". However, Eddington also thought deeply about more basic subjects, particularly the way in which science is done, and was of the opinion that much of physics is subjective, insofar as we necessarily filter data about the external world through our human-based senses. Hence the oft-repeated quote: "To put the conclusion crudely — the stuff of the world is mind-stuff". The purpose of the present section is to give a short and informal account of the man's views.

Eddington's influence was primarily through a series of non-technical books and his personal contacts with a series of great scientists who followed his lead. These include Dirac, Hoyle and McCrea. In the preceding section, we noted that while it is possible to add an arbitrary number of extra dimensions to relativity as an exercise in mathematics, we need to use the fundamental constants to identify their relevance to physics. (We are here talking primarily about the speed of light $c$, the gravitational constant $G$ and Planck's constant of action $h$, which on division by $2\pi$ also provides the quantum of spin angular momentum.) To appreciate Eddington's legacy, we note that his writings contain the first logical account of the large dimensionless numbers which occur in cosmology, thereby presaging what Dirac would later formalize as the Large Numbers Hypothesis. This consists basically in the assertion that large numbers of order $10^{40}$ are in fact equal, which leads among other consequences to the expectation that $G$ is variable over the age of the universe (see Wesson 1978; this possibility is now frequently discussed in the context of field theory in $N > 4$ dimensions). One also finds in Eddington's works some very insightful, if controversial, comments about the so-called fundamental constants. These appear to have influenced Hoyle, who argued that the $c^2$ in the common relativistic expression $(c^2 t^2 - x^2 - y^2 - z^2)$ should not be there, because "there is no more logical reason for using a different time unit than there would be for measuring $x$, $y$, $z$ in different units". The same influence seems to have acted on McCrea, who regarded $c$, $G$ and $h$ as "conversion constants and nothing more". These comments are in agreement with the view advanced in Section 3.2 above, namely that the fundamental constants are parameters which can be used to change the physical units of material quantities to lengths, enabling them to be given a geometrical description.

There is a corollary of this view which is pertinent to several modern versions of higher-dimensional physics. Whatever the size of the manifold, the equations of the related physics are homogeneous in their physical units $(M, L, T)$ so they can always be regarded as equalities involving dimensionless parameters. It was Dicke who clarified much work in variable–'constant' cosmology by emphasizing that physics basically consists of the comparison of dimensionless parameters at different points in the manifold. In other words, we measure numbers, often a physical quantity divided by its corresponding unit. When a dimensionless number which varies with time is decomposed into its dimensionful parts, there is bound to be controversy about which of those parts are variable and in what way. It is this ambiguity which has led some workers to discount modern theories of higher-dimensional physics in which the coupling 'constants' are variable. While Eddington did not explicitly develop variable–'constant' cosmologies, like the ones later proposed by Dicke, Dirac and Hoyle, he did lay the foundation for them by questioning the origin and nature of the fundamental constants. For this and other novel aspects of his writings, he was severely criticized by physicists and philosophers in the 1930s and 1940s (see Chapter 1). His current status is somewhat higher, because of the development of consistent 'variable' cosmologies in the intervening years. However, there is an interesting question of psychology involved here, which goes back to the age of the Greeks.

Plato tells us of an artisan whose products are the result of experience and skill and meet with the praise of his public for many years. However, in later times he suddenly produces a work which is stridently opposed to tradition and he incurs widespread criticism. Has the artisan suffered some delusion, or has he broken through to an art form so novel that his pedestrian-minded customers cannot appreciate or understand it?

Eddington spent the first part of his academic career doing well-regarded research on stars and other aspects of conventional astronomy. He then showed great insight and mathematical ability in his study of the then-new subject of general relativity. In his later years, however, he delved into the arcane topic of the dimensionless numbers of physics, attempting to derive them from an approach which combined elements of pure reason and mathematics. This approach figures significantly in his book *Relativity Theory of Protons and Electrons* (1936), and in the much-studied posthumous volume *Fundamental Theory* (1946). The approach fits naturally into his philosophy of science, which argues that many results in physics are the result of how we do science, rather than direct discoveries about the external world (which, however, he admitted). Jeffreys succeeded Eddington to the Plumian Chair at Cambridge, but was a modest man more interested in geophysics and the formation of the solar system than the speculative subject of cosmology. Nevertheless, he developed what at the time was a fundamental approach to the theory of probability, and applied his skills to a statistical analysis of Eddington's results. The conclusion was surprising: according to Jeffreys' analysis of the uncertainties in the underlying data which Eddington had used to construct his account of the basic physical parameters, the results agreed with the data better than they ought to have done. This raised the suspicion that Eddington had 'cooked' the results. This author spent the summer of 1970 in Cambridge, having written (during the preceding summer break from undergraduate studies at the University of London) a paper on geophysics which appealed to Jeffreys. We discussed, among other things, the status of Eddington's results. Jeffreys had great respect for Eddington's abilities, but was of the opinion that his predecessor had unwittingly put hidden elements into his approach which accounted for their unreasonable degree of perfection. The writer pointed out that there was another possible explanation: that Eddington was in fact right in his belief that the results of physics were derivable from first principles, and that his approach was compatible with a more profound theory which is yet to come.

## 3.4    Campbell and His Theorem

Whatever the form of a new theory which unifies gravity with the forces of particle physics, there is a consensus that it will involve extra dimensions. In Section 3.2, we considered mainly the 5D approach, which by the modern names of induced-matter and membrane theory is essentially old Kaluza–Klein theory without the stifling condition of compactification. The latter, wherein the extra dimension is 'rolled up' to a very small size, answers the question of why we do not 'see' the fifth dimension. However, an equally valid answer to this is that we are constrained to live close to a hypersurface, like an observer who walks across the surface of the Earth without being directly aware of what lies beneath his feet. In this interpretation, 5D general relativity must be regarded as a kind of new standard. It is the simplest extension of Einstein's theory, and is widely viewed as the low-energy limit of more sophisticated theories which accommodate the internal symmetry groups of particle physics, as in 10D supersymmetry, 11D supergravity and 26D string theory. There is, though, no sacrosanct value of the dimensionality $N$. It has to be chosen with a view to what physics is to be explained. (In this regard, St. Kalitzin many years ago considered $N \to \infty$.) All this understood, however, there is a practical issue which needs to be addressed and is common to all higher-$N$ theories: How do we embed a space of dimension $N$ in one of dimension $(N+1)$? This is of particular relevance to the embedding of 4D Einstein theory in 5D Kaluza–Klein theory. We will consider this issue in the present section, under the rubric of Campbell's theorem. While it is central and apparently simple, it turns out to have an unusual history with some novel implications.

John Edward Campbell was a professor of mathematics at Oxford whose book *A Course of Differential Geometry* was published posthumously in 1926. The book is basically a set of lecture notes on the algebraic properties of ND Riemannian manifolds, and the question of embeddings is treated in the latter part (notably Chapters 12 and 14).

However, what is nowadays called Campbell's theorem is only sketched. He had intended to add a chapter dealing with the relation between abstract spaces and Einstein's theory of general relativity (which was then a recent addition to physics), but died before he could complete it. The book was compiled with the aid of Campbell's colleague, E.B. Elliot, but while accurate is certainly incomplete.

The problem of embedding an ND (pseudo-) Riemannian manifold in a Ricci-flat space of one higher dimension was taken up again by Magaard. He essentially proved the theorem in his Ph.D. thesis of 1963. This and subsequent extensions of the theorem have been discussed by Seahra and Wesson (2003), who start from the Gauss–Codazzi equations and consider an alternative proof which can be applied to the induced-matter and membrane theories mentioned above.

The recognition of Campbell's theorem by physicists can be attributed largely to the work of Tavakol and coworkers. They wrote a series of articles in the mid-1990s which showed a connection between Campbell's theorem and a large body of earlier results by Wesson and coworkers (later reviewed by Wesson 2006). The latter group had been using 5D geometry as originally introduced by Kaluza and Klein to give a firm basis to the aforementioned idea of Einstein, who wished to transpose the "base-wood" of the right-hand side of his field equations into the "marble" of the left-hand side. That an effective or induced 4D energy-momentum tensor $T_{\alpha\beta}$ can be obtained from a 5D geometrical object such as the Ricci tensor $R_{AB}$ is evident from a consideration of the number of degrees of freedom involved in the problem. The only requirement is that the 5D metric tensor be left general, and not be restricted by artificial constraints such as the 'cylinder' condition imposed by Kaluza and Klein (no dependence on derivatives with respect to the extra coordinate). Given a 5D line element, it is then merely a question of algebra to show that the Ricci equations $R_{AB} = 0$ contain the ones $G_{\alpha\beta} = (8\pi G / c^4) T_{\alpha\beta}$ named after Einstein. Many exact solutions of $R_{AB} = 0$ are now known (see Wesson 2007 for a catalog). Of these, special mention should be made of the 'standard' 5D cosmological ones due to Ponce de Leon, and the 1-body and other solutions in the 'canonical' coordinates introduced by Mashhoon et al. It says something about the divide between physics and mathematics, that the connection between these solutions and Campbell's theorem was only made later, by the aforementioned work of Tavakol et al. Incidentally, these workers also pointed out the implications of the theorem for lower-dimensional ($N < 4$) gravity, which some researchers believe to be relevant to the quantization of this force.

Campbell's theorem is a local embedding theorem. It provides a formal, mathematical basis for embedding Einstein-like equations for ND in Ricci-like equations for $(N + 1)$D, since the number of degrees of freedom of the former set is less than (or equal to) that of the latter set. But it cannot be pushed towards solving problems which are the domain of (more difficult) global embeddings. This implies that Campbell's theorem should not be applied to initial-value problems or situations involving singularities. It is a modest (but still very useful) result, whose main implication is that we can gain a better understanding of matter in 4D by looking at the field equations in 5D. It also has the wider implication that, given the physics in a certain manifold, we can always derive the corresponding physics in a manifold of plus-or-minus one dimension. In other words, Campbell's theorem provides a kind of ladder which enables us to go up or down between manifolds of different dimensionality.

## 3.5    Conclusion

Dimensions are a delightful subject with which to dally, but we should remind ourselves that they need the cold scrutiny of common sense to be useful. This means, among other things, that we should have physical identifications for the extra coordinates, in order to understand the implications of their associated dimensions. In 4D, the presence of the fundamental constants $G$ and $c$ in Einstein's equations tells us that we are dealing with a relativistic theory of gravity. In 5D, we have seen that the extra coordinate can profitably be related to rest mass, either as the Schwarzschild radius or the Compton wavelength, in the classical and quantum domains respectively. This implies that the fifth dimension is a scalar field, which is presumably related to the Higgs field by which particles acquire mass in quantum field theory. This interpretation depends on a judicial use of the fundamental constants, and owes much to the work of Eddington, who delved deeply into the meanings of the equations of physics. Our usage of dimensions also owes something to Campbell, whose theorem in its modern form shows how to go between manifolds whose dimensionality differs by one. Our conclusion is that to be of practical importance, we need to ascribe the appropriate physical labels to the coordinates and the spaces, something which requires not only algebra but also skill.

## References

Barrow, J.D., 1981. Quart. J. Roy. Astron. Soc. <u>22</u>, 388.

Barrow, J.D., Tipler, F.J., 1986. The Anthropic Principle. Oxford University Press, New York.

Campbell, J.E., 1926. A Course of Differential Geometry. Clarendon Press, Oxford.

Eddington, A.S., 1935. New Pathways in Science. Cambridge University Press, Cambridge.

Eddington, A.S., 1936. Relativity Theory of Protons and Electrons. Cambridge University Press, Cambridge.

Eddington, A.S., 1939. The Philosophy of Physical Science. Cambridge University Press, Cambridge.

Eddington, A.S., 1946. Fundamental Theory. Cambridge University Press, Cambridge.

Green, M.B., Schwarz, J.H., Witten, E., 1987. Superstring Theory. Cambridge University Press, Cambridge.

Gubser, S.S., Lykken, J.D., 2004. Strings, Branes and Extra Dimensions. World Scientific, Singapore.

Halpern, P., 2004. The Great Beyond: Higher Dimensions, Parallel Universes, and the Extraordinary Search for a Theory of Everything. J. Wiley, Hoboken, N. J.

Kilmister, C.W., 1994. Eddington's Search for a Fundamental Theory. Cambridge University Press, Cambridge.

McCrea, W.H., Rees, M.J. (eds.), 1983. Phil. Trans. Roy. Soc. (London) A 310, 209.

Petley, B.W., 1985. The Fundamental Constants and the Frontier of Measurement. Hilger, Bristol.

Price, K., French, S. (eds.), 2004. Arthur Stanley Eddington: Interdisciplinary Perspectives. Centre for Research in the Arts, Humanities and Social Sciences (10–11 March), Cambridge.

Seahra, S.S., Wesson, P.S., 2003. Class Quant. Grav. 20, 1321.

Szabo, R.J., 2004. An Introduction to String Theory and D-Brane Dynamics. World Scientific, Singapore.

Wesson, P.S., 1978. Cosmology and Geophysics. Hilger/Oxford University Press, New York.

Wesson, P.S., 1992. Space Science Rev. 59, 365.

Wesson, P.S., 2006. Five-Dimensional Physics: Classical and Quantum Consequences of Kaluza–Klein Cosmology. World Scientific, Singapore.

Wesson, P.S., 2007. Space-Time-Matter: Modern Higher-Dimensional Cosmology, 2nd edn. World Scientific, Singapore.

West, P., 1986. Introduction to Supersymmetry and Supergravity. World Scientific, Singapore.

# Chapter 4

# TIME AS AN ILLUSION

## 4.1 Introduction

The concept of time can have different meanings for the physicist, the philosopher and the average person. In this chapter, we will widen the discussion of the preceding one, and attempt to arrive at an understanding of time which is broad-based and flexible.

In doing this, it will be necessary to debunk certain myths about time, and to clarify statements that have been made about it. Certainly, time has been a puzzling concept throughout history. For example, Newton in his Principia (Scholium I), stated that "Absolute, true and mathematical time, of itself, and from its own nature, flows equably without relation to anything external, and by another name is called duration." This sentence is often quoted in the literature, and is widely regarded as being in opposition to the nature of time as embodied later in relativity. However, prior to that sentence, Newton also wrote about time and space that "… the common people conceive these quantities under no other notions but from the relation they bear to sensible objects." Thus Newton was aware that the "common" people in the 1700s held a view of time and other physical concepts which was essentially the same as the one used by Einstein,

Minkowski, Poincaré and others in the 1900s as the basis for relativity.

As a property of relativity, it is unquestionably true that the time $t$ can be considered as a physical dimension, on the same basis as our measures $(xyz)$ of three-dimensional space. It was Minkowski who argued in a famous speech that time should be welded to space to form spacetime. The result is a hybrid measure of separation, or interval, commonly called the Minkowski metric. It is the basis of quantum mechanics. By extension to curved as opposed to flat spacetime, we obtain a more complicated expression for the interval, which is the basis of cosmology. In both applications, the numerical value of the interval is given by a kind of super-Pythagorean sum, in which the squares of elements are added together (though with a sign difference between the time and space parts to indicate their different natures). The time part involves the product of $t$ with the speed of light $c$, which essentially transforms the 'distance' along the time axis to a length $ct$. Due to this, the interval is also a measure of which points are (or are not) in contact via the exchange of photons. Those particles with real interval can be in contact, while those with imaginary interval cannot be in contact.

This way of presenting Minkowski spacetime is conventional and familiar. However, it has a corollary which is not so familiar: particles with zero interval are coincident in 4D. Einstein realized this, and it is the basis of his definition of simultaneity. But it is not a situation which most people find easy to picture, so they decompose 4D spacetime into 3D space and 1D time, and visualize a photon propagating through $xyz$ over time $t$. Eddington, the noted contemporary of Einstein, also appreciated the subjective nature of the situation just described, and went on to argue that much of what is called objective in physics is in fact subjective or invented. The speed of light was also commented on later by a few deep thinkers such as McCrea and Hoyle, who regarded it as a mere man-made constant (see Chapter 3). From the Eddington viewpoint, one can argue that the decomposition of 4D Minkowski spacetime into separate 3D and 1D parts is a subjective act, so that in effect the photon has been invented as a consequence of separating space and time.

Below, we will enlarge on the possibly subjective nature of physics, with an emphasis on the concept of time. We will in fact suggest that time is a subjective ordering device, used by humans to make sense of their world. Several workers have expressed this idea, including Einstein (1955 in Hoffmann 1972), Eddington (1928, 1939), Hoyle (1963, 1966), Ballard (1984) and Wesson (2001). We hope to show that this approach makes scientific sense, and from a common-day perspective has certain comforts.

Such an approach is, however, somewhat radical. So to motivate it, we wish to give a critique of other, more mainstream views. This will be short, because good reviews of the nature of time are available by many workers including Gold (1967), Davies (1974), Whitrow (1980), McCrea (1986), Hawking (1988), Landsberg (1989), Zeh (1992), Woodward (1995) and Halpern and Wesson (2006). We will discuss contending views of the nature of time in Section 4.2, introduce what seems to be a better approach in Section 4.3, and expand on the implications of this in Section 4.4. Although it is not essential, it will become apparent that our new approach to time is psychologically most productive when the world is taken to have more dimensions than the four of spacetime, in accordance with modern physics.

## 4.2   Physics and the Flow of Time

The idea that time flows from the past to the future, and that the reason for this has something to do with the natural world, has become endemic to philosophy and physics. However, this idea is suspect. We will in this section examine briefly the three ways in which the direction of time's 'arrow' is commonly connected with physical processes, and argue that they are all deficient. Quite apart from technical arguments, a little thought will show that a statement such as the "flow of time", despite being everyday usage, is close to nonsensical. For the phrase implies that time itself can be measured with respect to another quantity of the same kind. This might be given some rational basis in a multidimensional universe in which there is more than one time axis (see below); but the everyday usage implies measuring the change of a temporal quantity against itself, which is clearly a contradiction in terms. Such a sloppy use of words appears to be tolerated because there is a widespread belief that the subjective, unidirectional nature of time can be justified by more concrete, physical phenomena.

Entropy is a physical concept which figures in the laws of thermodynamics. Strictly speaking, it is a measure of the number of possible states of a physical system. But more specifically, it is a measure of the disorder in a system; and since disorder is observed to increase in most systems as they evolve, the growth of entropy is commonly taken as indicative of the passage of time. This connection was made by Eddington, who also commented on the inverse relationship between information and entropy (Eddington 1928, 1939). However, the connection has been carried to an unreasonable degree by some subsequent writers, who appear to believe that the passage of time is equivalent to the increase of entropy. That this is not so can be seen by a simple counter-argument: If it were true, each person could carry a badge that registered their entropy, and its measurement would correlate with the time on a local clock. This is clearly daft.

A more acceptable application of the notion of entropy might be found in the many-worlds interpretation of quantum mechanics. This was proposed by Everett (1957), and supported as physically reasonable by De Witt (1970). In it, microscopic systems bifurcate, and so define the direction of the future. In principle, this approach is viable. However, the theory would be better couched in terms of a universe with more than the four dimensions of spacetime; and interest in the idea of many worlds appears to have lapsed, because there is no known way to validate or disprove their existence.

Another physical basis for the passage of time which has been much discussed concerns the use of so-called retarded potentials in electromagnetism. The connection is somewhat indirect, but can be illustrated by a simple case where light propagates from one point to another. (This is what happens when humans apprehend things by the sense of sight, and is also how most information is transmitted by modern technology.) Let the signal be emitted at point P and observed at point O, where the distance between them is $d$ and is traversed at lightspeed $c$. Now Maxwell's equations, which govern the interaction, are symmetric in the time $t$. (We are assuming that the distance is small enough that ordinary three-dimensional space can be taken as Euclidean or flat.) However, in order to get the physics right, we have to use the electromagnetic potential not at time $t$ but at the retarded time $(t - d/c)$. This is, of course, the time 'corrected' for the travel lag from the point P of emission to the point O of observation. Such a procedure may appear logical; but it has been pointed out by many thinkers that it automatically introduces a time asymmetry into the problem (see Davies 1974 for an extensive review). The use of retarded potentials, while they agree with observations, is made even more puzzling by the fact that Maxwell's equations are equally valid if use is made instead of the 'advanced' potentials defined at $(t + d/c)$. In short, the underlying theory treats negative and positive increments of time on the same footing, but the real world appears to prefer the solutions where the past evolves to the future. Studies have been made of the symmetric case, called Wheeler/Feynman electrodynamics, where both retarded and advanced potentials are allowed. One argument for why we do not experience the signals corresponding to the advanced potentials is that due to Hoyle and Narlikar (1974). They reasoned that the unwanted signals would be absorbed in certain types of cosmological models, leaving us with a universe which is apparently asymmetric between the past and future. This explanation is controversial, insofar as it appeals to unverified aspects of the large-scale cosmos. On the small scale, it appears that the need for retarded potentials in electrodynamics leads to a locally-defined arrow of time; though whether this is due to objective physical reasons, or to some subjective bias on our part, remains obscure.

The big bang offers yet another way of accounting for the arrow of time. According to Einstein's theory of general relativity, everything we observe came into existence in a singularity at a specific epoch, which supernova data fix at approximately $13 \times 10^9$ years before the present (see elsewhere for more detailed discussions). This description is familiar to all, and carries with it the implication that the universe in a dynamical sense has a preferred direction of evolution. However, closer examination shows that it is really the recession of the galaxies from each other, rather than the big bang, which identifies the time-sense of the universe's evolution. This was understood by Bondi (1952), who was one of the founders with Gold and Hoyle of the steady-state theory. In it, matter is continuously created and condenses to form new galaxies, whose average density is thereby maintained even though the whole system is expanding. While no longer regarded as a practical cosmology, the steady-state theory shows that it is the motions of galaxies which essentially defines a preferred direction for time, rather than the (still poorly understood) processes by which they may have formed after the big bang. Let us, in fact, temporarily forget about the latter event, and consider an ensemble of gravitating galaxies. Then there are in principle only three modes of evolution: expansion, contraction and being static. The last can be ruled out, because it is widely acknowledged that such a state, even if it existed, would be unstable and tip into one of the other two modes. We are thus lead to the realization that if the arrow of time is dictated by the dynamical evolution of the universe, its sense is given *a priori* by a 50/50 choice analogous to flipping a cosmic coin. That is, there is no dynamical reason for believing that events should go forward rather than backwards in time. In addition to this, there is also the problem that there is no known physical process which can transfer a cosmic effect on a lengthscale of $10^{28}$ cm down to a human one of order $10^2$ cm. In order to circumvent this objection, it has been suggested that the humanly-perceived arrow of time is connected instead to smaller-scale astrophysics, such as the nucleosynthesis of elements that determines the evolution of the Sun. This process might, via the notion of entropy as discussed above, be connected to geophysical effects on the Earth, and so to the biology of its human inhabitants. But it is really obvious, when we pick apart the argument, that there is no discernable link between the mechanics of the evolving universe and the sense of the passage of time which is experienced by people.

The preceding issues, to do with entropy, electrodynamics and cosmology, have the unfortunate smell of speculation. Dispassionate thought reveals little convincing connection between the time coordinate used in physics and the concept of age as used in human biology. We can certainly imagine possible connections between physical and human time, as for example in *Einstein's Dreams* by

Lightman (1993). There, the effects of relativity such as time dilation are described in sociological contexts. But, there is a large gap between the fluid manner in which time can be manipulated by the novelist and the rigid transformations of time permitted to the physicist. Indeed, while the physicist may be able to handle the "$t$" symbol in his equations with dexterity, he looks clumsy and strained when he attempts to extend his theories to the practicality of everyday existence. That is why the sayings about time by physicists mainly languish in obscurity, while those by philosophers and others have wider usage.

In the latter category, we can consider the statement of the philosophical novelist Marcel Proust: "The world was not created at the beginning of time. The world is created every day." This appears to dismiss the big bang, and by implication other parts of physics, as irrelevant to the human experience of time. However, it is more rewarding to consider statements like the foregoing as pointed challenges to the physicist. To be specific: Is there a view of "time" which is compatible with the rather narrow usage of the word in physics, and yet in agreement with the many ways in which the concept is experienced by people?

## 4.3   Time as a Subjective Ordering Device

The differing roles which time plays in physics and everyday life has led some workers to the conclusion that it is a subjective concept. Let us consider the following quotes:

Einstein (as reported by Hoffman): "For us believing physicists the distinction between past, present and future is only an illusion, even if a stubborn one."

Eddington: "General scientific considerations, favour the view that our feeling of the going on of time is a sensory impression; that is to say, it is as closely connected with stimuli from the physical world as the sensation of light is. Just as certain physical disturbances entering the brain cells via the optic nerves occasion the sensation of light, so a change of entropy ... occasions the sensation of time succession, the moment of greater entropy being felt to be the later."

Hoyle: "All moments of time exist together." "There is no such thing as 'waiting' for the future." "It could be that when we make subjective judgments we're using connections that are non-local... there is a division, the world divides into two, into two completely disparate stacks of pigeon holes."

Ballard: "Think of the world as a simultaneous structure. Everything that's ever happened, all the events that will ever happen, are taking place together." "It's possible to imagine that everything is happening at once, all the events 'past' and 'future' which constitute the universe are taking place together. Perhaps our sense of time is a primitive mental structure that we inherited from our less intelligent forbears."

The preceding four opinions about time have an uncanny similarity, given that they apparently originate independently of each other. However, they are all compatible with Eddington's view of science, wherein certain concepts of physics are not so much discovered as invented (see Wesson 2000 for a short review). The subjective nature of time is also compatible with certain views of particle physics and cosmology, wherein several worlds exist alongside each other (Everett 1957, De Witt 1970, Penrose 1989, Wesson 2006, Petkov 2007). It is important to realize that there need not be anything mystical about this approach. For example, Hoyle considers a 4D world of the usual type with time and space coordinates $t$ and $xyz$ which define a surface $\phi(t, xyz) = C$. Here $C$ is a parameter which defines a subset of points in the world. Changing $C$ changes the subset, and "We could be said to live our lives through changes of $C$." In other words, the life of a person can be regarded as the consequence of some mechanism which picks out sets of events for him to experience.

What such a mechanism might be is obscure. Hoyle speculated that the mechanism might involve known physical fields such as electromagnetism, which is the basis of human brain functions. It might plausibly involve quantum phenomena, amplified to macroscopic levels by the brain in the manner envisaged by Penrose (1989). However, while the precise mechanism is unknown, some progress can be made in a general way by noting that Hoyle's $C$-equation above is an example of what in relativity is known as a hypersurface. This is the relation one obtains when one cuts through a higher-dimensional manifold, defining thereby the usual 4D world we know as spacetime. It is in fact quite feasible that the Minkowski spacetime of our local experience is just a slice through a world of more than 4 dimensions.

We will investigate this in some detail in the next chapter. For now, suffice it to state that higher dimensions are the currently popular way to unify gravity with the interactions of particle physics, and that reviews of the subject are readily available (e.g., Wesson 2006 from the physical side and Petkov 2007 from the philosophical side). Since we are here mainly interested in the concept of time, let us concentrate on one exact solution of the theory for the simplest case when there is only a single extra dimension. (See Wesson 2007 for a compendium of higher-dimensional solutions including the one examined here.) Let us augment the time ($t$) and the coordinates of Euclidean space ($xyz$) by an extra length ($l$). Then by solving the analog of Einstein's equations of general relativity in 5D, the interval between two nearby points can be written

$$dS^2 = l^2 dt^2 - l^2 \exp i(\omega t + k_x x) dx^2 - l^2 \exp i(\omega t + k_y y) dy^2$$
$$- l^2 \exp i(\omega t + k_z z) dz^2 + L^2 dl^2. \tag{4.1}$$

Here $\omega$ is a frequency, $k_x$ etc. are wave numbers and $L$ measures the size of the extra dimension. This equation, while it may look complicated, has some very informative aspects: (a) it describes a wave, in which parts of what are commonly called space can come into and go out of existence; (b) it can be transformed by a change of coordinates to a flat manifold, so what looks like a space with structure is equivalent to one that is featureless; (c) the signature is $+ - - - +$, so the extra coordinate acts like a second time. These properties allow of some inferences relevant to the present discussion: (i) even ordinary 3D space can be ephemeral; (ii) a space may have structure which is not intrinsic but a result of how it is described; (iii) there is no unique way to identify time.

This last property is striking. It means that in grand-unified theories for the forces of physics, the definition of time may be ambiguous. This classical result confirms the inference from quantum theory, where the statistical interaction of particles can lead to thermodynamic arrows of time for different parts of the universe which are different or even opposed (Schulman 1997, 2000). It should be noted that the existence of more than one 'time' is not confined to 5D relativity, but also occurs in other $N$-dimensional accounts such as string theory (Bars et al. 1999). Indeed, there can in principle be many time-like coordinates in an $N$-dimensional metric.

In addition, the definition of time may be altered even in the standard 4D version of general relativity by a coordinate transformation. (This in quantum field theory is frequently called a gauge choice.) The reason is that Einstein's field equations are set up in terms of tensors, in order to ensure their applicability to any system of coordinates. This property, called covariance, is widely regarded as essential for any modern theory of physics (see Section 5.5). However, if we wish to have equations which are valid irrespective of how we choose the coordinates, then we perforce have to accept the fact that time and space are malleable. Indeed, covariance even allows us to mix the time and space labels. Given the principle of covariance, it is not hard to see why physicists have abandoned the unique time label of Newton, and replaced it by the ambiguous one of Einstein.

We are led to the realization that the concept of time is as much a puzzle to the physicist as it is to the philosopher. Amusingly, the average person in the street probably feels more comfortable about the issue than those who attempt to analyse it. However, it is plausible that time in its different guises is a device used by people to organize their existence, and as such is at least partially subjective in character.

## 4.4     Mathematics and Reality

In the foregoing, we saw that several deep thinkers have arrived independently at a somewhat intriguing view of time. To paraphrase them: time is a stubborn illusion (Einstein), connected with human sensory impressions (Eddington), so that all moments of time exist together (Hoyle), with the division between past and future merely a holdover from our primitive ancestors (Ballard). Perhaps the most trenchant opinion is that of Hoyle (1966), who summarizes the situation thus: "There's one thing quite certain in this business. The idea of time as a steady progression from past to future is wrong. I know very well we feel this way about it subjectively. But we're all victims of a confidence trick. If there's one thing we can be sure about in physics, it is that all times exist with equal reality."

This view of time can be put on a physical basis. We imagine that each person's experiences are a subset of points in spacetime, defined technically by a hypersurface in a higher-dimensional world, and that a person's life is represented by the evolution of this hypersurface. This is admittedly difficult to visualize. But we can think of existence as a vast ocean whose parts are all connected, but across which a wave runs, its breaking crest precipitating our experiences.

A mathematical model for a wave in five dimensions was actually considered in the preceding section as equation (4.1). It should be noted that there is nothing very special about the dimensionality, and that it is unclear how many dimensions are required to adequately explain all of known physics. The important thing is that if we set the interval to zero, to define a world whose parts are connected in higher dimensions, then we necessarily obtain the hypersurface which defines experience in the lower-dimensional world. It is interesting to note that the behaviour of that hypersurface depends critically on the number of plus and minus signs in the metric (i.e. on the signature). In the canonical extension of Einstein's theory of general relativity from four to five dimensions, the hypersurface has two possible behaviours. Let us express the hypersurface generally as a length, which depends on the interval of spacetime $s$, or equivalently on what physicists call the proper time (which is the time of everyday existence corrected to account for things like the motion). Then the two possible behaviours for the hypersurface may be written

$$l = l_o \exp(s/L) \quad \text{and} \quad l = l_o \exp(is/L). \tag{4.2}$$

Here $l_o$ is a fiducial value of the extra coordinate, $L$ is the length which defines the size of the fifth dimension, and $s$ is the aforementioned interval or proper time. The two noted behaviours describe, respectively, a growing mode and an oscillating mode. The difference between the two modes depends on the signature of the metric, and is indicated by the absence or presence of $i \equiv \sqrt{-1}$ in the usual manner. So far, the analysis follows the basic idea about experience due to Hoyle but expressed in the language of hypersurfaces as discussed by Wesson (see Hoyle and Hoyle 1963,

Wesson 2006). However, it is possible to go further, and extend the analysis into the metaphysical domain for those so inclined. This by virtue of a change from the growing mode to the oscillatory mode, with the identification of the former with a person's material life and the latter with a person's spiritual life. That is, we obtain a simple model wherein existence is described by a hypersurface in a higher-dimensional world, with two modes of which one is growing and is identified with corporeal life, while one is wave-like and is identified with the soul, the two modes separated by an event which is commonly called death.

Whether one believes in a model like this which straddles physics and spirituality is up to the individual. (In this regard, the author is steadfastly neutral.) However, it is remarkable that such a model can even be formulated, bridging as it does realms of experience which have traditionally been viewed as immutably separate. Even if one stops part way through the above analysis, it is clear that the concept of time may well be an illusion. This in itself should be sufficient to comfort those who fear death, which should rather be viewed as a phase change than an endpoint.

## 4.5    Conclusion

Time is an exceptionally puzzling thing, because people experience it in different ways. It can be formalized, using the speed of light, as a coordinate on par with the coordinates of ordinary three-dimensional space. But while spacetime is an effective tool for the physicist, this treatment of time seems sterile to the average person, and does not explain the origin of time as a concept. There are shortcomings in purely physical explanations of time and its apparent flow, be they from entropy, many-worlds, electromagnetism or the big bang. Such things seem too abstract and remote to adequately explain the individual's everyday experience of time. Hence the suggestion that time is a subjective ordering device, invented by the human mind to make sense of its perceived world.

This idea, while not mainstream, has occurred to several thinkers. These include the novelist/philosopher Proust, the physicists/ astronomers Einstein, Eddington and Hoyle, and the futurist Ballard. It is noteworthy that the idea appears to have its genesis independently with these people. And while basically psychological in nature, it is compatible with certain approaches in physics, notably Penrose's suggestion that the human brain may be a kind of amplification organ for turning tiny, quantum-mechanical effects into measurable, macroscopic ones. The idea of time as an ordering device was given a basis in the physics of relativity by Hoyle, who however only sketched the issue, arguing that the movement of a hypersurface would effectively provide a model for the progress of a person's life. This approach can be considerably developed, as outlined above, if we assume that the experience-interface is related to a 4D hypersurface in a 5(or higher)D world. Then it is possible to write down an equation for the hypersurface, which can have an evolutionary and an oscillatory phase, which might (if a person is so inclined) be identified with the materialistic and spiritual modes of existence. Perhaps more importantly, in this 5D approach, the interval (or 'separation') between points is zero, so all of the events in the world are in (5D) causal contact. In other words, everything is occurring simultaneously.

That this picture may be difficult to visualize just bolsters the need for something like the concept of time, which can organize simultaneous sense data into a comprehensible order.

Time, viewed in this manner, is akin to the three measures of ordinary space, at least insofar as how the brain works. Humans have binocular vision, which enables them to judge distances. This is an evolutionary, biological trait. Certain other hunting animals, like wolves, share it. By comparison, a rabbit has eyes set into the sides of its head, so while it can react well to an image that might pose a threat, it cannot judge distance well. But even a human with good vision finds it increasingly difficult to judge the relative positions of objects at great distance: the world takes on a two-dimensional appearance, like a photograph, or a landscape painting. In the latter, a good artist will use differing degrees of shade and detail to give an impression of distance, as for example when depicting a series of hills and valleys which recede to the horizon. Likewise, the human brain uses subtle clues to do with illumination and resolution to form an opinion about the relative spacing of objects at a distance. This process is learned, and not perfectly understood by physiologists and psychologists; but is of course essential to the adequate functioning of an adult person in his or her environment. Astronomers have long been aware of the pitfalls of trying to assess the distances of remote objects. In the past, they measured offsets in longitude and latitude by means of two angles indicated by the telescope, called right ascension and declination. But they had no way of directly measuring the distances along the line of sight, and so referred to their essentially 2D maps as being drawn on the surface of an imaginary surface called the celestial sphere. Given such a limited way of mapping, it was very hard to decide if two galaxies seen close together on the sky were physically close or by chance juxtaposed along the line of sight. In lieu of a direct method of distance determination, astronomers fell back on probability arguments to decide (say) if two galaxies near to each other on a photographic plate were really tied together by gravity, or merely the result of a coincidental proximity in 2D while being widely separated in 3D. This situation changed drastically when technological advances made it easier to measure the redshifts of galaxies, since the redshift of a source could be connected via Hubble's law to the physical distance along the line of sight. Nowadays, by combining angular measurements for longitude and latitude with redshifts for outward distance, astronomers have fairly good 3D maps of the distribution of galaxies in deep space.

In effect, astronomers have managed to replace the photograph (which is essentially 2D) by the hologram (which provides information in 3D). However, whether this is done for a cluster of galaxies or a family portrait, the process of evaluating distance is a relatively complicated one. The human brain evaluates 3D separations routinely, and we are not usually aware of any conscious effort in doing so. But this apparently mundane process is also a complicated one. If we take it that the concept of time is similar to the concept of space, it is hardly surprising that the human brain has evolved its own subtle way of handling 'separations' along the time axis of existence.

Then the idea of time as a kind of subjective ordering device, by which we make sense of a simultaneous world, appears quite natural.

## References

Ballard, J.G., 1984. Myths of the Near Future. Triad-Panther, London.

Bars, I., Deliduman, C., Minic, D., 1999. Phys. Rev. D 59, 125004.

Bondi, H., 1952. Cosmology. Cambridge University Press, Cambridge.

Davies, P.C.W., 1974. The Physics of Time Asymmetry. University of California Press, Berkeley.

De Witt, B.S., 1970. Phys. Today 23 (9), 30.

Eddington, A.S., 1928. The Nature of the Physical World. Cambridge University Press, Cambridge.

Eddington, A.S., 1939. The Philosophy of Physical Science. Cambridge University Press, Cambridge.

Everett, H., 1957. Rev. Mod. Phy. 29, 454.

Gold, T. (ed.), 1967. The Nature of Time. Cornell University Press, Ithaca, N.Y.

Halpern, P., Wesson, P.S., 2006. Brave New Universe: Illuminating the Darkest Secrets of the Cosmos. J. Henry, Washington, D.C.

Hawking, S.W., 1988. A Brief History of Time. Bantam Press, New York.

Hoffmann, B., 1972. Albert Einstein, Creator and Rebel. New American Lib., New York.

Hoyle, F., 1966. October the First is Too Late. Fawcett-Crest, Greenwich, Conn.

Hoyle, F., Hoyle, G., 1963. Fifth Planet. Heinemann, London.

Hoyle, F., Narlikar, J.V., 1974. Action at a Distance in Physics and Cosmology. Freeman, San Francisco.

Landsberg, P.T., 1989. *In* Physics in the Making (Sarlemijn, A., Sparnaay, M. J., eds.). Elsevier, Amsterdam, 131.

Lightman, A., 1993. Einstein's Dreams. Random House, New York.

McCrea, W.H., 1986. Quart. J. Royal Astron. Soc. 27, 137.

Petkov, V. (ed.), 2007. Relativity and the Dimensionality of the World. Springer, Berlin.

Penrose, R., 1989. The Emperor's New Mind. Oxford University Press, Oxford.

Schulman, L.S., 1997. Time's Arrow and Quantum Measurement. Cambridge University Press, Cambridge.

Schulman, L.S., 2000. Phys. Rev. Lett. 85, 897.

Wesson, P.S., 2000. Observatory 120 (1154), 59. Ibid., 2001, 121 (1161), 82.

Wesson, P.S., 2006. Five-Dimensional Physics: Classical and Quantum Consequences of Kaluza–Klein Cosmology. World Scientific, Singapore.

Wesson, P.S., 2007. Space-Time-Matter: Modern Higher-Dimensional Cosmology, 2nd edn. World Scientific, Singapore.

Whitrow, G.J., 1980. The Natural Philosophy of Time. Oxford University Press, Oxford.

Woodward, J.F., 1995. Found. Phys. Lett. 8, 1.

Zeh, H.-D., 1992. The Physical Basis of Time. Springer, Berlin.

# Chapter 5

# THE NATURE OF MATTER

## 5.1    Introduction

The puzzle of how to define matter was highlighted by Eddington, who knocked on the top of his desk, and then observed that while it could bear his weight, it was in fact almost entirely empty space.

Certainly an atom is almost entirely empty space: it consists of widely-spaced, miniscule electrons orbiting a dense nucleus. But even the latter, including the nucleus of the hydrogen atom which is just a proton, is largely devoid of what most people understand by the word "matter". Modern physics avoids the contradiction between something and nothing by appeal to the concept of the vacuum, which loosely speaking is a state that is not matter but yet contains energy. In Einstein's general theory of relativity, the energy density of the vacuum is measured by the cosmological constant $\Lambda$. Recent observations of the universe indicate a significant, positive value of $\Lambda$. In terms of density, the universe appears to be dominated by dark energy of $\Lambda$-type, to the extent of about 74%. Most of the rest is exotic dark matter, meaning that it is believed to consist of particles as opposed to fields, but with unfamiliar properties. Only a few percent of the universe is now believed to consist of the material seen in stars and galaxies, or what used to be termed ordinary matter

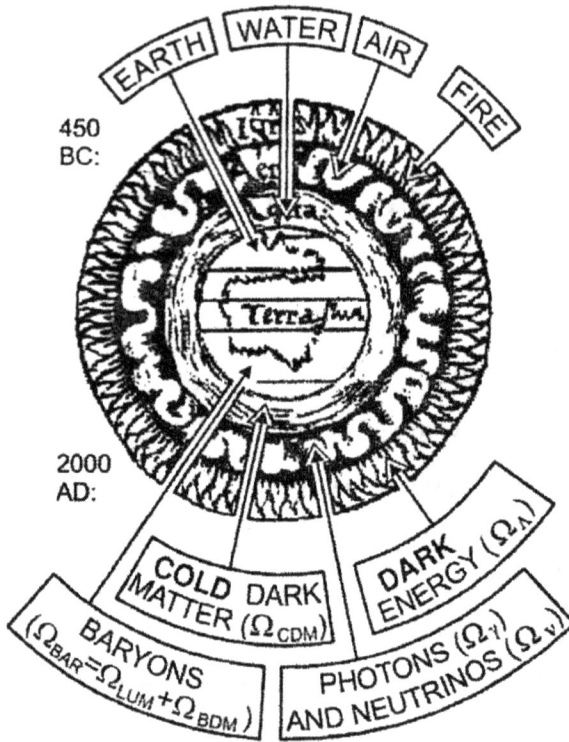

**Figure 5.1.** The components of the world according to ancient philosophers and modern astrophysicists. The current view has it that the universe is dominated by dark energy, to which is added some kind of cold dark matter, with a sprinkling of the kind of baryons (heavy particles) found in stars and galaxies, with the whole bathed in a dilute sea of photons and neutrinos. The only commonality between the old and new views is the division into four elements, and even that is subjective.

(Figure 5.1). Of course, it is always possible to convert between an energy and a mass by using Einstein's formula and the square of the speed of light. This and other considerations bring up the question of whether it is useful to distinguish between dark energy, dark matter and ordinary matter. The answer is in the affirmative, because these components of the universe have distinct modes of behaviour. In fact, the nature of matter is largely determined by its behaviour.

In what follows, we will therefore not attempt to give any universal meaning to the word "matter", realizing that (like many terms in physics) it is a flexible concept. Rather, we will attempt to understand matter by examining its behaviour in certain circumstances.

## 5.2    Properties of Matter

For largely historical reasons, it is common practice to define matter in terms of three parameters: the density $\rho$, the pressure $p$ and the temperature $T$. Of these, the first two are more mechanical in nature than the last. They can be determined either in the laboratory, or within the context of an established theory for some remote object that cannot be examined at close quarters. (For example, in Newtonian gravity the density of an interstellar cloud and its associated gravitational potential are related by Poisson's equation, which we will study in Section 5.5 below.) By contrast, the temperature $T$ is not macroscopic in character but microscopic, and whether it is measured locally or remotely, intrinsically involves atomic or quantum physics. (For example, in Maxwellian gas theory the energy associated with one degree of motion in ordinary space is $kT/2$, where $k$ is Boltzmann's constant, which basically provides a way of going between the mechanical and thermodynamical concepts of energy.) A relation between the three parameters $\rho$, $p$ and $T$ is known as an equation of state. In practice, it may not be possible to measure $T$ directly, or it may be only poorly known. This is often the case in astrophysics and cosmology. Therefore, we frequently write an equation of state in the form $p = p(\rho)$, a relation between the pressure and the density.

Some equations of state are relatively simple. This is because the parameters $\rho$ and $p$ are treated as scalars. We expect, of course, that the macroscopic density should be isotropic in the three directions of ordinary space, since the inertial rest masses of the particles that make up the sample have this property, both by experiment, and by construction for most theories. The pressure, however, could conceivably be different in the three directions of ordinary space, reflecting microscopic anisotropies to do, for example, with temperature gradients. But the case where the pressure is isotropic is fairly common. For this case, and neglecting other effects which might cause anisotropy such as viscosity or the presence of a magnetic field, the fluid is described by only the two parameters $\rho$ and $p$. This is called a perfect fluid. Due to its simplicity, it is the source usually assumed for solutions of Einstein's equations of general relativity (see below). A sceptic might point out that these equations are almost impossible to solve unless we assume a perfect-fluid source, but we leave this aside for now.

There are several equations of state for a perfect fluid which are of interest:

(1) $p = 0$ describes dust. There are no microscopic interactions between the particles of the fluid, so equivalently the temperature is zero.

(2) $p = \rho c^2 / 3$ describes electromagnetic radiation. In a particle description, the photons have zero rest mass but finite energies and

momenta, and move at lightspeed $c$. This is also the limiting equation of state for particles with finite rest masses, which however move with speeds so close to $c$ that they are ultrarelativistic and resemble photons.

(3) $p = \alpha\rho c^2$ describes an isothermal fluid. This is the state where the particles have the same temperature throughout the sample, and formally includes the case $\alpha = 1/3$ noted above for photons. The case $\alpha = 1$ is a limiting one, where sound waves travel at the speed of light, a limit which cannot be exceeded given standard definitions of causality. The case $\alpha = 1$ is frequently referred to as "stiff" matter; but matter in the laboratory and in most stars obeys $\alpha \ll 1$.

(4) $p = -\rho c^2$ describes the vacuum of general relativity. This is a unique state for Einstein's theory, because the cosmological constant $\Lambda$ is a simple constant. The connection between $\Lambda$ and the hypothetical vacuum fluid is provided by the fact that $\Lambda$ can be moved from the left-hand to the right-hand side of Einstein's equations, and interpreted not as a dynamical term but as a source term. The precise connection is via a density and pressure, in the absence of ordinary matter, with the values $\rho = \Lambda c^2 / 8\pi G$ and $p = -\Lambda c^4 / 8\pi G$. Here $\Lambda$ is taken to have the physical dimensions of an inverse length squared, and $G$ is the gravitational constant. This trick — of reinterpreting $\Lambda$ as a measure of vacuum energy — is in widespread use. But it should be noted that the $8\pi G$ in the denominators of $\rho$ and $p$ exactly cancels the $8\pi G$ which is in the numerator of the energy-momentum tensor, which forms the source term in general relativity (see elsewhere). This implies that the interpretation of the cosmological constant in

terms of a vacuum fluid may be contrived. Also, the invariance properties of the energy-momentum tensor necessarily imply that a positive (inertial) density for the vacuum means a negative pressure. So if $\Lambda > 0$ as indicated by astrophysical data, the vacuum has a negative effective pressure; and considered as a fluid, the vacuum is therefore of a different type from those studied in the laboratory.

The above four examples of perfect-fluid equations of state actually cover most of what is needed to study modern astrophysics and cosmology. However, it should be noted that the case of $\rho = 0$, $p = 0$ has not been mentioned. Indeed, we have gone through the above presentation of what are well-known results partly to illustrate an important point: the old-fashioned idea of emptiness with $\rho = 0 = p$ is not of much use to modern physics.

There are good reasons, from both observation and theory, for this. Telescope data, combined with classical cosmology as based on general relativity, show that $\Lambda$ is finite and positive, so the 'ground state' of the universe is not that defined by the traditional notion of emptiness. This is confirmed by accelerator data, which on the basis of standard quantum field theory show that particles have large vacuum fields, which can in an approximation also be expressed in terms of a microscopic value of $\Lambda$. (Though as discussed in Chapter 2, there is controversy about its magnitude.) From the theoretical side, it is also not really surprising that the empty state with $\rho = 0 = p$ is irrelevant. For this state follows from the field equations when the spacetime is the purely Minkowski one of special relativity. That is, the old idea of emptiness corresponds to a situation in which the metric coefficients or potentials are perfectly smooth and have magnitudes exactly equal to one, with no fields of any kind due to the presence of real objects. This is clearly unrealistic. Even if we imagine an empty universe into which we introduce one particle, for example as a base for an observer, we automatically deform spacetime away from its pure Minkowski form. In short, the idea of a completely empty universe is an abstraction.

Mach was the first to clearly realize that matter and space should not be treated as separate entities, but as parts of an organic whole (Figure 5.2). Mach motivated Einstein in the development of general relativity, though it is widely held that the ideas of the former are not completely incorporated into the equations of the latter. Einstein's approach was taken up by Eddington, who fully appreciated that cosmology required a vacuum ground-state related to the cosmological constant. (The fact that Einstein later renounced the cosmological constant appears to have been the only major scientific issue on which the two men differed.) Mach's influence on Einstein and Eddington was not restricted to philosophical considerations about matter and spacetime. Indeed, Mach's views about how masses interact in space led indirectly to the three principles on which general relativity is based, and so to the detailed models for astrophysics and cosmology which are today's staples. Consider, as an illustration, the simple case in which a large object (such as the Sun) interacts with a small one (such as the Earth). This interaction should, as far as we are able, be described in terms which are independent of how we choose coordinates. This implies a mathematical formulation in terms of

**Figure 5.2.** Mach (1838–1916) was a largely home-schooled physicist, whose belief that local masses are influenced by remote matter in the universe motivated Einstein to formulate general relativity.

tensors (see elsewhere), and is formalized as the Covariance Principle. Even given this, however, there is an ambiguity in the problem of the interaction of masses. This becomes clearer when we realize that the mass which causes the gravitational field is logically distinct from the mass which responds to the force and measures energy (see Figure 5.3). Einstein removed this ambiguity by stating their similarity explicitly, in the Equivalence Principle. Lastly, we still need a prescription for how a test object reacts to the gravitational field of a larger body (e.g., how the Earth's orbit is determined by the

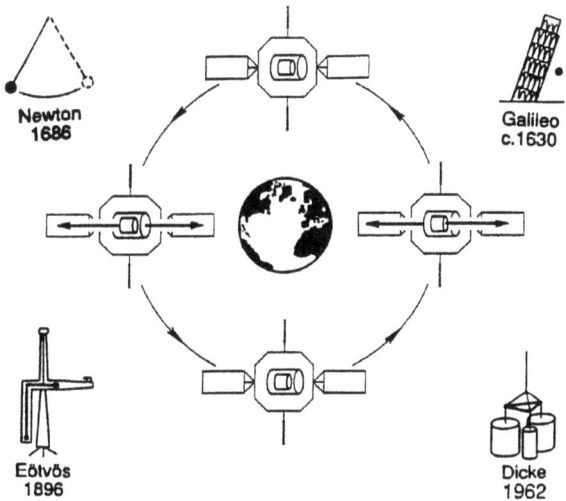

**Figure 5.3.** It is a postulate of Einstein's theory of general relativity that the kind of mass which is associated with the gravity of an object is proportional to the kind of mass which measures the energy content of the object. We in fact traditionally assume this by using the same symbol for the gravitational mass and the inertial mass. However, this basic property of mass needs to be tested. This figure illustrates the methods which have been used through time to test the proportionality of the two types of mass. From top right, these include the putative dropping of objects from the leaning tower of Pisa by Galileo, the motion of a pendulum as studied by Newton, and the development of the torsion balance by Eotvos and Dicke. The centre ring motif around the Earth illustrates a planned experiment called the Satellite Test of the Equivalence Principle. In this, test masses aboard a spacecraft feel the effect of the Earth's gravitational field, so the equivalence of gravitational and inertial mass can be verified to an accuracy of one part in $10^{18}$.

Sun). For this, we go back to the Fermat rule for the 'shortest' path, as discussed in Chapter 1, and embody it in the Geodesic Principle. The three Principles just out-lined form the basis of Einstein's general theory of relativity.

There are, however, other theories which respect the foregoing trio of principles. Einstein's theory is, after all, primarily an account of the gravitational aspects of matter. If we are to understand the other interactions shown by matter — electromagnetism and the strong and weak forces of particles — we expect to have to widen the theory beyond general relativity. Nowadays, the consensus view is that to unify the gravitational interaction of matter with the short-range forces of particles requires an extension of spacetime to a manifold with more than four dimensions. These issues are discussed at other, more appropriate places (for reviews see the books by Mach 1893, Eddington 1939, Barbour and Pfister 1995, Wesson 2006). Here we draw the conclusion that the properties of matter are an integral part of the equations in which they appear; and that in a comparison between theory and experiment, we are obliged to adopt a very broad view of what is simplistically called "matter".

## 5.3    Creating Matter

In the laboratory, the best-studied example of the creation of matter is the production of an electron/positron pair from an electromagnetic field. This process is well understood, because the energy involved is relatively low (the rest-mass energy of the electron is close to 0.5 MeV). The creation of other types of particles is less well understood, because the energies are higher. However, it is already apparent from the example of pair creation that we are not dealing with some miraculous process; but rather with the conversion of energy from one form to another, in accordance with the laws of conservation. (Most notably, in the electron/positron case, the total electric charge is conserved.) By comparison, the big bang is a different and somewhat anomalous event. In it, all of the matter in the universe — whether in the form of the rest masses of particles or the energies of fields — is supposed to come into existence at some instant. It is a spontaneous event, with no prehistory and no analyzable physics.

Researchers appear to either love or hate the big bang. In a way, it is a beautiful thing, in that a singularity in Einstein's equations cannot by its nature be mathematically or physically traversed, and provides a kind of natural beginning for the rest of science. In a different way, however, it is abhorrent just because it is not like events in the rest of physics, and rebuffs any attempt at analyzing the origin of matter.

Insofar as it is possible to examine the big bang dispassionately, it is clear that a lot of the negative attitudes about it is due to a misunderstanding of its nature. It is not an explosion of conventional type, say due to a bomb, which sends out shrapnel from a uniquely-defined point in ordinary three-dimensional space (we pointed this out before in Chapter 2). Rather, consider numerous bombs, distributed endlessly throughout space, which all detonate at the same instant. There is no focus in 3D, since the explosion occupies all of ordinary space. The event is only uniquely defined in time.

Cosmological models based on general relativity are not, however, uniquely defined just by the fact that they begin in a big bang. In fact, they are not uniquely defined even if we know the time which has elapsed since that event. Figure 5.4 shows the allowed

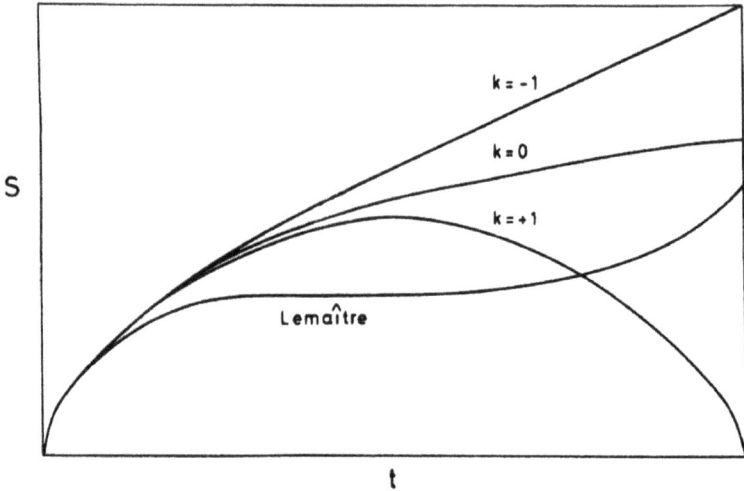

**Figure 5.4.** The behaviour of the scale factor $S$ with time $t$ in standard models of the universe. The scale factor is proportional to the separation of any two galaxies that take part in the expansion of the universe (not ones that are gravitationally bound to each other or are members of the same cluster). This diagram is sometimes given with the vertical axis labeled as a radius, but this is misleading because it suggests a boundary and there is no such thing in these models. There are three basic types of behaviour, depending on the value of a constant $k$ that appears in the theory of general relativity. For $k = +1$, the model expands to a maximum and then collapses. For $k = 0$, the model expands but slows and becomes static for very large times. For $k = -1$, the model expands throughout its history. Also shown is another type of behaviour named after its discoverer Lemaitre. This model has $k = +1$, but is modified by the introduction of a positive cosmological constant (taken to be zero in the other curves shown). This constant is equivalent to a repulsive force that counteracts gravity. For this case, the model expands, becomes nearly static, and then expands again. All of these models start in a big bang and have infinite density at time zero. And all have significant problems in comparison with observation.

behaviours for the scale of the expanding universe as it depends on the time. In the suite of allowed models, modern data from a variety of sources indicate that our universe is of a particularly simple type (namely, that where the curvature of ordinary three-dimensional space has the normalized value $k$ of zero). Furthermore, recent observations of supernova stars inform us that the elapsed time since the big bang (if there was one) is approximately $13 \times 10^9$ years.

This is a convenient number, being about three times the age of the Earth. Alternatively, the age of the universe is slightly less than three times the age of the Sun. The latter, and the myriads of other stars like it, provide a means of checking the age of the visible matter in the universe. For as stars produce energy, it is diluted by the expansion of the galaxies and the redshift effect, so that there is an intergalactic field of radiation whose intensity depends on the time which has elapsed since the galaxies (and their stars) formed. This background field owes its origin to processes like thermonuclear fusion, and should not be confused with the microwave background, which is believed to be the cooled-down radiation from the pregalactic fireball. (For a detailed discussion, see Section 2.2.) For those who do not believe in the big bang, the origin of the microwave background has always been a point of contention. But the intensity of the ordinary, *star-produced* radiation rests on undisputed physics (Overduin and Wesson 2008). And a match between the observed intensity and models of cosmology based on general relativity gives the age of the galaxies. This is close to the above-quoted figure. Thus the existence of the galaxies, and their accumulated starlight, show us that something special happened in the universe, and that it did so about 13 Gyr ago.

Irrespective of whether we believe that most of the matter in the universe was created in a big bang, that event comes out of Einstein's theory of general relativity. And that theory has much more to say about the evolution of matter than just the possibility of an initial explosion. As in Newtonian gravity, in Einstein's theory matter is governed by certain laws. One set of these are the equations of motion, which tell how a fluid with density $\rho$ and pressure $p$ will evolve under the influence of gravity. In relativity, the strength of gravity is still governed by the constant $G$ introduced by Newton, but now in conjunction with the speed of light $c$. Let us consider a fluid which has the same properties about some point in 3D space (isotropy), and is also the same when that point is moved (homogeneity). In short, the fluid is uniform. For ease conceptual, let us also assume that the fluid is perfect, in the sense defined above. Since we are dealing with relativity, the underlying scaffold for measurement is spacetime, wherein the three axes of ordinary space are linked to the one 'axis' of time (see Chapter 4). This means that in general that there will be four equations of motion (one for each axis), though the fourth or temporal one will have a different character. In fact, the 'motion' along the time axis in relativity is connected to the energy of a test particle, in a way analogous to how the motions along the spatial axes are connected to the linear momenta. For a fluid, the equivalent equation of 'motion' for the temporal axis turns out to be connected to the conservation of matter. We may skip the details of the analysis (which are to be found in standard texts for 4D relativity, and for higher dimensions in Wesson 2006). Also, because we are considering a fluid which is uniform in ordinary 3D space, the three spatial equations of motion are identical and reduce to one. To write it down, it is convenient to refer things to a local but arbitrary centre of coordinates, and to measure the distance from this to some other point by the radius $R$. (This symbol does not imply the existence of a physical boundary, or a physically-special origin, since our fluid is uniform, and by postulate does not possess either thing.) We choose to use a dot to denote the derivative with respect to time, so the acceleration is $\ddot{R}$. Then with all our symbols defined, we can write down the equations of motion for a fluid in general relativity as just two relations. One is the standard formula for the acceleration of the fluid, and the other is the formula which expresses the conservation of its matter. These relations are:

$$\ddot{R} = \frac{-4\pi G R}{3c^2}(3p + \rho c^2) \tag{5.1}$$

$$\dot{\rho}c^2 = \frac{-3\dot{R}}{R}(p + \rho c^2). \tag{5.2}$$

These two modest-looking equations represent a fount of physics, as we will see.

The first equation says that the acceleration due to gravity is towards the centre (hence the minus sign). It may be shown that it is actually proportional to the inverse square of $R$, if the right-hand side of (5.1) is recast by appropriately defining the mass of the fluid interior to radius $R$ by an integral over this and the properties of matter (see below). The situation is therefore similar to that in Newtonian gravity. However, in the latter, the speed of light does not appear, and the strength of the source is just the density $\rho$. By contrast, in Einsteinian gravity as described by (5.1), the speed of light plays a crucial role and the strength of the source is the combination $(3p + \rho c^2)$. It makes sense that the density is augmented by the pressure, because this measures the motions of the particles which make up the fluid, and so measures their kinetic energies. That is, the total source is the sum of the rest masses of the particles as measured by the density $\rho$, and the mass equivalent of their kinetic energies as measured by the pressure $p$. We see that the ratio $p / \rho c^2$, which is tiny in laboratory physics, can have significant effects in astrophysics. The sum $(3p + \rho c^2)$ is called the gravitational energy density.

The second equation above says that the density of the fluid goes down if its properties are standard and if there is expansion ($\dot{R} > 0$). Also, in general the density and size of a portion of the fluid vary together, when the pressure is negligible, in such a way as to keep constant the appropriately defined mass (see below). This is similar to the conservation of mass in Newtonian theory, which is usually formalized by the equation of continuity. This is modified in Einsteinian gravity as in (5.2), where the important factor is the combination $(p + \rho c^2)$. It makes sense that this is the governing factor, because we recall from above that in general relativity the equation of state of the vacuum as measured by the cosmological constant is $p = -\rho c^2 = -\Lambda c^4 / 8\pi G$. Therefore, the vacuum state in

Einstein's theory has $(p + \rho c^2) = 0$, and is stable. Other states, which will usually be mixtures of vacuum and ordinary matter, will evolve with time in accordance with (5.2). The sum $(p + \rho c^2)$ is called the inertial energy density.

The reader may wish to tinker with equations (5.1) and (5.2) above, to explore the implications of matter creation allowed by the various combinations of $\dot{R}$ and $p$ (both of which may be positive or negative in principle). It is already apparent that $p$ is a phenomenological parameter, in the sense that it is a macroscopic, classical label for microscopic and possibly quantum processes for which details may be wanting. There are, however, other more conventional processes for which the effective value of $p$ can be negative, such as turbulence. The combination $\dot{R} > 0$, $p < 0$ can occur in the early universe, when particles may be trying to attract each other by short-range forces but cannot overcome the global expansion. This is a particularly interesting case, because it has been used to quantify a model universe with appealing philosophical properties: it exists forever in a Minkowski state with no motion, no density and no pressure; then an event of possible quantum nature (such as tunneling) upsets the equilibrium, causing expansion to start, while the pressure goes negative, leading to the density undergoing a sharp surge to positive values. The model goes through an era with a hot or radiation-like equation of state, and eventually settles into a dust-like mode with conventional properties (Figure 5.5). This model, and others like it, has been studied in some detail. For those so inclined, it has the advantage of replacing the big bang by a big blip.

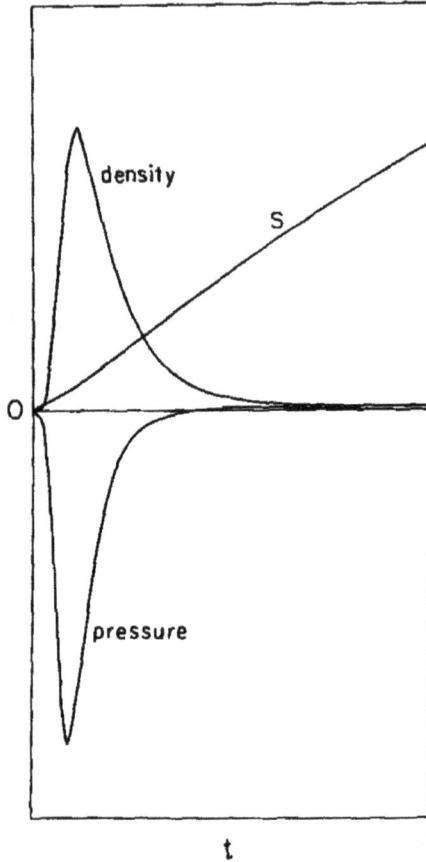

t

**Figure 5.5.** This illustrates the kind of universe that has a big 'blip' rather than a big bang. The average distance between free galaxies is described by a scale factor $S$ that depends on the time $t$. But unlike the infinite density and pressure at $t = 0$ of the big bang, here the density rises to a peak because the pressure is allowed to go into a negative trough. Thereafter, the matter goes through a hot or radiation-like phase, and then settles down into a model of conventional type. Models like this, which avoid the big bang but still obey the equations of general relativity, have been studied by Bonnor and Wesson. They are interesting, because if so desired they can be preceded by a flat, empty phase (Minkowski space) where there was nothing — until some quantum 'kick' started things off.

Mass — as opposed to density — is a difficult quantity to deal with in general relativity. This may seem odd. But from a technical viewpoint it is understandable, because the Einstein field equations are local in nature, relating the curvature of spacetime to the density and other properties of matter, all at a chosen point in space and time. By contrast, mass involves the summation of the properties of particles over a finite region, and is therefore a secondary concept in which various factors of definition come into play. There are, indeed, about a half-dozen different definitions of mass in use for general relativity. This immediately leads us to suspect that the concept of mass is at least partially subjective, an inference which is borne out by detailed investigation.

For our present purposes, we will concentrate on one specially convenient definition, which is in widespread use because it is of practical importance for astrophysics. In that context, many systems are spherically symmetric in the three dimensions of ordinary space. (That is, all of the important parameters depend only on a suitably-defined distance $R$ from a chosen origin, and are independent of the angles that define distances in the two directions orthogonal to the radius.) Also, the material can often be well approximated by a perfect fluid, so only the symbols $\rho$ and $p$ used before need to be considered. When this type of problem is set up in accordance with Einstein's field equations of general relativity, the 3D spherical symmetry leads to considerable simplification. The problem can be cast in the form of four, second-order, partial differential equations. Now, it is a well-used ploy to reduce equations which are second order in the derivatives to ones which are first order (and therefore easier to tackle), by the introduction of a new quantity. This has to be chosen carefully, of course, if it is to help towards a solution. But in our case, it turns out that the new quantity we need is not only mathematically convenient but also physically relevant; it is effectively the mass. Let us denote this by the symbol $M$, where though we understand that it is the mass interior to radius $R$ in a fluid which may be expanding or contracting at rate $\dot{R}$. (We are here not concerned with the intrinsic differences in distances due to 3D curvature, because they are taken care of in the definitions of $R$ and $M$, and we wish to obtain a relation which is not only mathematically correct but also physically informative.) Our problem now has the status not of four, second-order equations but of five, first-order equations, of which one is essentially a definition for the mass $M$. Of these five equations, one is particularly important in regard to the question of the origin of matter. It reads

$$\dot{M}c^2 = -4\pi p R^2 \dot{R}. \tag{5.3}$$

This can be understood as expressing a balance of power (or the rate of doing work). The right-hand side involves the pressure ($p$) acting over the area of a spherical shell ($4\pi R^2$), so forming a force; which is multiplied by a velocity ($\dot{R}$) to give what every engineer recognizes as a power. The left-hand side is just the rate of change of the energy inside the corresponding surface, expressed using the mass ($M$) and the speed of light ($c$) in accordance with the usual Einstein formula. In other words, (5.3) is a statement about the conservation of energy through time.

If we wish to explain the origin of matter in accordance with the classical laws of physics, equations (5.3) and (5.2) show that we need the pressure to be negative if the universe is expanding. Then the mass as given by (5.3) or the density as given by (5.2) can increase. This process is entirely compatible with the standard theory of general relativity.

Continuous creation, by contrast, is a process that logically requires new physics outside of Einstein's theory. It refers to the creation of particles from apparently empty space, and was part of an attempt which was made in the years 1950–1970 to expand 4D relativity. The main motivation for this was the wish to widen the group of invariances on which gravitation is based. Straight general relativity ensures by the use of tensors that its equations are valid irrespective of how we change the coordinates with which we describe things. It is not, however, invariant under changes in the length scales with which we describe things. (Elementary examples of changes in coordinates and scales are the shift from Cartesians to spherical polars, and the shift from centimeters to inches, respectively.) Some researchers have taken the view that Einstein's equations for gravitation, like Maxwell's equations for electromagnetism, ought to be invariant under changes of both types. Dirac termed this co-covariance, as a way of indicating that it is an extension of the usual invariance under a change of coordinates. Other authors, like Hoyle and Narlikar and Canuto et al., termed the requirement simply scale invariance. The various versions of 4D general relativity proposed by these and other workers all involved the possibility that the number of particles in a given region of space could change slowly over cosmic time. In Hoyle's approach, this was connected with the operation of a new entity called a "C-field" ("C" for Creation). Unfortunately, his and the other versions of this theory ran into problems with observational astrophysics. A major obstacle is that observations indicate that most galaxies formed at one particular epoch, which is hardly compatible with the spread of ages expected from continuous creation.

Extended versions of general relativity which are based on 4D spacetime are in any case now regarded as obsolete. This because they offer no clear way to unify gravitation with the interactions of particles, something which is widely regarded as best approached through extra dimensions. There is an extensive literature on $N (> 4) D$ relativity (see Wesson 2006 and elsewhere). Since we are here discussing the nature of matter, we note that the 5D theory is a direct extension of the 4D one, in which a new field is added that affects particle masses. The new field is scalar in type, as opposed to the tensor one of Einstein gravity and the vector one of Maxwell electrodynamics. However, the addition of a scalar potential also means that four other potentials of 'mixed' type appear, and these are commonly identified with those of electromagnetism. The theory is thus a classical unified account of gravity, electromagnetism and a scalar/mass interaction. The corresponding quantum theory is an account of the spin-2 graviton (the hypothetical particle which mediates gravity), the spin-1 photon and a spin-0 scalaron (some aspects of it are discussed in Section 2.4). However, the quantum embodiment of the theory is not completely worked out. It is believed, though, that the scalar field of the classical theory is related to the Higgs field of quantum field theory, which is responsible in effect for boosting the masses of the elementary particles from zero to their observed values.

The creation of an electron/positron pair from the electromagnetic field, as outlined at the beginning of this section, can be used as a model to study the corresponding processes in the gravitational field and the scalar field (if this exists). However, these are quantum-mechanical processes. The details of such processes cannot be captured by a classical field theory, whether it uses 4D spacetime or a 5D manifold. In astrophysics and cosmology, we are bound to use relations like (5.1)–(5.3) above. The last of these, we recall, involves a definition for the mass. This is acceptable, because it is based on the field equations of general relativity and includes terms we expect to find, such as contributions from the rest mass, the mass-equivalent of the kinetic energy and the curvature of the spacetime. But it is a definition, nonetheless. And as such, it is at least partially subjective in nature, in the sense that we could have chosen some other one. (Several alternatives actually exist, as noted before.) Of course, the definition which is embodied in (5.3) is justified by its utility. But even so, we are obliged to consider the possibility that the pressure can be negative, at least if we wish to analyse the creation of matter rather than merely accepting its existence as a consequence of the big bang. If the pressure can be negative, is it sensible to consider the possibility that other properties of matter might also be negative?

## 5.4    Negative Mass?

In a way, negative mass has been a part of physics for ages, because gravitational binding energy is negative, and on dividing by $c^2$ defines what is formally a negative mass. But this is a cheap answer to the question; and in this section we wish to take a brief look at the more significant possibility that the mass of a discrete object like a particle can be negative.

This is not a silly question. It cannot be immediately dismissed by the fact that astrophysics has not revealed any objects with negative mass. For an object with a negative gravitational mass would repel other objects, instead of attracting them. It should be recalled that even in Newtonian gravity, the interaction between a large mass $M$ and a test mass $m$ that are separated by distance $r$ involves an acceleration $a$ given by $ma = GMm / r^2$. (The usage of the same symbol $m$ on both sides of this equation actually involves the Equivalence Principle mentioned above, since it allows us to identify what are logically distinct types of mass related to inertia and gravity.) Cancelling the $m$ symbol, we obtain $a = GM / r^2$. So a negative-mass object repels all test particles, irrespective of whether the latter are themselves positive or negative in nature. In a universe dominated by negative-mass objects, they would not congregate to form galaxies and the other structures we observe. Indeed, a medium consisting of negative-mass particles would tend by its very nature to be dispersed, and therefore difficult to detect.

A more technical analysis of the feasibility of negative mass was give by Bonnor (1989). He examined the postulates and laws which characterize our knowledge of gravitation, and came to a somewhat surprising conclusion: there is no way to rule out negative mass, at least from a theoretical standpoint.

A clue as to why this should be is contained in modern theories of gravitation, in which 4D spacetime is extended by an extra dimension related to rest mass (Wesson 2008). In the version of 5D general relativity known as space-time-matter theory, the mass of an object is effectively measured by a parameter which has the nature of a length, like a coordinate. This, of itself, can be either positive or negative. However, the quantity which defines measurable aspects of the geometry depends not on this mass length directly, but on the square of it. That is, there is a kind of invariance or symmetry involved. It is of the same kind as those involving the space and time parts of the theory. These latter can be easily codified: $P$ denotes the invariance under reflection in space or what is technically called parity, while $T$ denotes invariance under time reversal. (There is also $C$ to denote charge invariance for those applications of the theory which include electric charge.) We see that 5D relativity may involve a new kind of symmetry for the mass, namely $M$. In the real world, processes at the particle level obey a combined symmetry, in accordance with the CPT theorem. It can be conjectured that the reason we do not directly observe negative-mass particles is that a kind of CPTM theorem is in effect.

## 5.5    Manipulating Matter

Properties of matter such as the density, pressure and mass have to be allowed to take on wide ranges of values if they are to be useful in modern physics. The fact that we are willing to stretch the meanings of theses parameters far beyond their historical ranges, rather than introduce new ones, tells us that physics has a kind of philosophical inertia. It is more acceptable to bend the framework of known physical theory, and manipulate the meanings of its symbols, than to step outside what has been established.

A typical example of this philosophy is provided by the decay of the neutron to produce a proton and an electron. Observations of the energies of the particles concerned before and after decay showed a mismatch. But the principle of conservation of energy was sufficiently established that instead of abandoning or modifying it, the unexplained energy difference was attributed to a new particle, the neutrino.

Another example is provided by the genesis of special relativity, which is really attributable to the reluctance of physicists to abandon the inviolate nature of the speed of light $c$ in frames that move past each other with constant velocities. The immutability of $c$, as realized by Einstein, Poincaré and others, meant a corresponding downgrade in how observers regard time and space. The latter cannot be concrete precepts in the manner of Newton, but must instead be concepts that are malleable. Along with this change in perception, it is also necessary to throw away the vast and intricate framework which had been built up about the hypothetical medium that supported light waves, the aether. The history of the aether is, in itself, fascinating (Whittaker 1910). It is a lesson to walk through the dungeons of a large library and pick out volumes on physics for the years 1850–1900. Most of them are preoccupied with the aether, and have long discussions dedicated to problems which nowadays would be considered laughable. (For example, whether the jagged tops of the Earth's mountain ranges, causing friction as they plough through the aether, would not result in the planet slowing in its orbit and spiraling in to the Sun.) While it is common to regard the aether as the biggest folly of physics, at least it was jettisoned in favour of simpler ideas when it became necessary.

There is, however, no guarantee that physics will not paint itself into another logical corner as it evolves. Indeed, some physicists are of the opinion that the subject during its present phase of rapid development is not only racking up successes, but also producing an uncomfortable number of paradoxes, as discussed in Chapter 2. But the consensus appears to believe that the course of physics is set fair, at least for a while. This is largely because the theories we currently have, and the symbols they involve, have a good degree of "stretch" left in them.

The flexibility of modern physical theory is actually remarkable. This may be appreciated by considering the case of matter and gravitation.

Newtonian gravity is encapsulated by one simple equation named after Poisson. It reads

$$\nabla^2\phi = 4\pi G\rho. \tag{5.4}$$

This relates the gravitational potential $\phi$ to the density of ordinary matter $\rho$ (we are assuming that the pressure $p$ is negligible, so $p/\rho c^2 \ll 1$). The second-order derivatives with respect to the three directions of ordinary space are combined via $\nabla^2 \equiv \partial^2 / \partial x^2$ $+ \partial^2 / \partial y^2 + \partial^2 / \partial z^2$ in Cartesian coordinates. A test particle of mass $m$ in the gravitational field defined by $\phi = \phi(xyz)$ has an acceleration given by $\partial \phi / \partial x$, with similar expressions along the other two axes. The acceleration, when multiplied by $m$, gives the force on the test particle. This has the familiar form of Newton's inverse-square law, when all of the matter is concentrated in one spot and the rest of space is empty. In this case, (5.4) reads just $\nabla^2 \phi = 0$, the equation which is named after Laplace. This is arguably the simplest yet most profound equation in physics. It appears in all branches of the subject, and in gravitation has many more solutions than the elementary inverse-square one of Newton. If a picture is worth a thousand words, then an equation must be worth a million such. The short statement "$\nabla^2 \phi = 0$" opens to the physicist a wealth of possibilities.

The equations of Poisson and Laplace apply to many situations, but they are ones in which things do not evolve significantly with time. In the event there is noticeable evolution along the time axis of spacetime, we need to add in the temporal coordinate $ct$ in the manner suggested by Minkowski and Einstein. Technically, this involves the Minkowski tensor of spacetime, which can be thought of as a $4 \times 4$ matrix, with nonzero components only along the diagonal: $\eta_{\alpha\beta} \equiv (+1, -1, -1, -1)$. However, while we will need this below, here we merely need to extend the $\nabla^2$ operator introduced above by the addition of a time component with the appropriate sign. Then Laplace's equation becomes $\square^2 \phi = 0$, where $\square^2 \equiv \partial^2 / c^2 \partial t^2 - \partial^2 / \partial x^2$ $- \partial^2 / \partial y^2 - \partial^2 / \partial z^2$. This, like its time-independent predecessor, has many applications. One of these is to waves, and for that reason $\square^2\phi = 0$ is sometimes called the wave equation.

Einsteinian gravity can also be stated in one equation, though this is somewhat deceptive in that what is written on one line is actually shorthand for a set of relations (Einstein 1950). We considered these field equations before, but for convenience repeat them here:

$$G_{\alpha\beta} = (8\pi G / c^4)T_{\alpha\beta}. \qquad (5.5)$$

This relates the gravitational potentials involved in the Einstein tensor $G_{\alpha\beta}$ to the properties of matter encoded in the energy-momentum tensor $T_{\alpha\beta}$ (we are assuming that the subscripts $\alpha$ and $\beta$ run over the coordinates of spacetime, where it is convenient to label $x^0 = ct$ for the time, and $x^{123} = xyz$ or some equivalent system for ordinary 3D space). Both $G_{\alpha\beta}$ and $T_{\alpha\beta}$ are tensors, so (5.5) holds in all systems of coordinates. They can be thought of as $4 \times 4$ arrays of elements, with the important proviso that they are symmetric. This means that the elements on one side of the diagonal are the mirror image of those on the other side. Now for any such array in $N$ dimensions, the total number of elements is $N^2$ and the number along the diagonal is just $N$. The number in one of the two, off-diagonal sectors is thus $(N^2 - N)/2$. This plus the elements along the diagonal is the total number of independent components, which is $N(N+1)/2$. For 4D spacetime, this is 10. Therefore, Einstein's field equations (5.5) are actually a set of 10 relations.

It is natural to ask at this stage if we are not making things unnecessarily complicated, in that we have gone from the single Poisson equation (5.4) to the 10 Einstein equations (5.5). The short answer to this is No. For the gravitational field is in reality more complicated than assumed previously (e.g., it involves gravitational waves whose torsional nature cannot be described by a scalar potential); and the properties of matter are more numerous than known before (e.g., they involve thermodynamical effects such as heat flow which go beyond what can be described by a simple density). It should also be recalled that the theory is actually simpler than it might otherwise have been because of the symmetry of its tensor ingredients. This can be traced to the symmetry of the basic potentials $g_{\alpha\beta}$, which in general relativity depend on the coordinates via $g_{\alpha\beta}(x^\gamma)$ instead of being constants as they were for special relativity (see above: $g_{\alpha\beta}$ replaces $\eta_{\alpha\beta}$). In fact, the potentials $g_{\alpha\beta}$ play a dual role in general relativity. Firstly, they allow us to broaden the old Pythagorean definition of the distance between two nearby points in space to the corresponding interval in spacetime, $ds^2 = g_{\alpha\beta}dx^\alpha dx^\beta$ (where a repeated index downstairs and upstairs indicates summation). Secondly, the derivatives of the potentials $g_{\alpha\beta}(x^\gamma)$ allow us to build up a select set of tensors which reflect the geometrical properties of the gravitational field. These are the Riemann–Christoffel tensor $R_{\alpha\beta\gamma\delta}$, the Ricci tensor $R_{\alpha\beta}$, the Ricci or scalar curvature $R$, and the Einstein tensor defined via $G_{\alpha\beta} \equiv R_{\alpha\beta} - (R/2)g_{\alpha\beta}$ which forms the left-hand side of the field equations (5.5). The last of these is constructed to have zero divergence (i.e., zero 'spread' in 4D). This matches the zero divergence of the energy-momentum tensor $T_{\alpha\beta}$, which is itself constructed in such a way from the properties of matter that we recover the conservation laws of physics. In short, the match between $G_{\alpha\beta}$ and $T_{\alpha\beta}$ expressed by (5.5) balances geometry with matter.

General relativity represented a monumental academic achievement, whose crux was the realization by Einstein that geometry could be used to represent the real world. The only comparable accomplishment in the history of physics was the realization by Newton that the force which caused an apple to fall to the ground was the same as the one which controlled the Moon in its orbit, and his formalization of the law of gravity and the attendant laws of motion. But despite the originality and scope of his theory of gravity, Einstein has been criticized for taking over a decade to go from special to general relativity. This kind of criticism usually comes from those who are only conversant with quantum theory, and is as lightweight as the particles they study. Einstein has also been chided for not being able, in his later years, to formulate a theory which unified gravity with the interactions of particles. This criticism has a (low) level of veracity. Einstein was supportive of the idea of extra dimensions and was familiar with the five-dimensional approaches of Kaluza (1921) and Klein (1926). However, the unification of gravitation and electromagnetism due to Kaluza was hobbled by discarding all derivatives with respect to the extra coordinate (the 'cylinder' condition); and the quantization of the electric charge due to Klein came at the expense of restricting to a circle the topology of the extra dimension ('compactification'). Furthermore, the theorem of Campbell (1926), which showed how to embed 4D in 5D, was hardly known and would have to wait till the 1990s for a meaningful application to physics (see Chapter 3). As it was, general relativity received scant attention compared to quantum theory until the 1960s, when Wheeler and others realized that it was essential to the study of astrophysics and cosmology.

In a space nearly empty of ordinary matter, such as the solar system, Einstein's equations (5.5) read just $G_{\alpha\beta} = 0$. These can be expressed in simpler form via the Ricci tensor introduced above, as $R_{\alpha\beta} = 0$. (We are here neglecting the cosmological constant, whose influence on the planets is negligible.) It is this form of the field equations which is verified by the classical tests of relativity. These and related tests of Einstein's theory have been extensively reviewed elsewhere (e.g., Will 1993). It is worth remarking, however, that while Eddington is often regarded as a physicist to whom thought took precedence over experiment, it was in fact he who was the main mover in verifying general relativity through solar eclipse observations. Data from these and other astrophysical sources verified Einstein over Newton, at least for empty space.

When matter is present, Einstein's theory is on a comparatively secure base because of the manner in which the field equations (5.5) are constructed. Let us consider matter in the form of a perfect fluid, whose energy density dominates the effects of pressure (see above). Then the only significant contribution to the energy-momentum tensor $T_{\alpha\beta}$ is through its zero-zero or time-time component, which is proportional to $\rho$. When the gravitational field is weak, it is straightforward to analyse the 10 equations in (5.5), and find that they reduce to just one relation (e.g., Rindler 2001). This is formally identical to Poisson's equation (5.4). In this way, we see that general relativity gives back the established physics of gravitating matter.

But what general relativity does *not* do is to tell us what matter *is*.

That it makes sense to ask this was clear to Einstein. He denigrated the crudity of the matter term in his equations (5.5) in contrast to the beauty of the gravitational term. Einstein's goal was to geometrize matter, in the same way as he had for the gravitational field. He also understood that the traditional division of the two concepts, in terms of the two sides of an equation, was convenient but artificial. In this, he was following Mach, who regarded a mass and its attendant gravitational field as symbiotic parts of the same thing. Unfortunately, as we noted above, Einstein failed in his search for a way to geometrize matter. In retrospect, this is doubly ironic: Einstein had the basic tools necessary for the job, but when it was finally carried out it was in ignorance of the great man's views.

Space-time-matter theory was formulated in 1992 as a means of explaining the origin of matter in a logically more sound way than the conventional big bang. In that year, the properties of matter of the standard cosmological models were derived from first principles; along with the corresponding properties for objects with spherical symmetry in ordinary 3D space ('solitons'). Also, the general expression for Einstein's energy-momentum tensor was written down in terms of pure geometry. These results were achieved by using a five-dimensional space of the sort proposed in the 1920s by Kaluza and Klein, but without the restrictions of the cylinder condition and compactification. The resulting 5D algebra is accordingly rich but heavy, so we content ourselves here with noting the more important developments in the theory (an historical review is given by Wesson

2008). These include the introduction of canonical coordinates by Mashhoon and others, which provided a way of geometrizing the mass of an individual particle; the application of Campbell's embedding theorem of differential geometry by Tavakol and others, which provided a mathematical basis for what had otherwise been proved by physics; and the discovery that the standard cosmological models of Ponce de Leon were flat in 5D though curved in 4D (Figure 5.6), implying that the big bang was a kind of artifact introduced by a poor choice of coordinates.

Matter as a result of geometry is a neat idea, but it should be implemented in a way which preserves Einstein's four-dimensional theory of relativity. As we noted previously, this involves 10 field equations. The extension to 5D, which is sometimes called induced-matter theory, involves 15 field equations. By analogy with the proven ones of Einstein's 4D theory, the equations (for the 5D theory) are couched in terms of the Ricci tensor, and read $R_{AB} = 0 (A, B = 0,123,4$ for time, space and the extra mass-related dimension). These equations may look as if they pertain to empty space; but actually they contain matter implicitly, as we will see.

Of the 15 just-noted equations, one is a wave equation for the extra potential ($g_{44} = \varepsilon \Phi^2$, where $\varepsilon = \pm 1$ indicates whether the extra dimension is spacelike or timelike, and the scalar field $\Phi$ can depend on the 4 coordinates of spacetime $x^\alpha$ plus the fifth one $x^4 = l$). Then there is a set of four relations which, even in the absence of electromagnetism, express the conservation of a kind of mass current. The remaining 10 relations are equivalent to Einstein's field equations

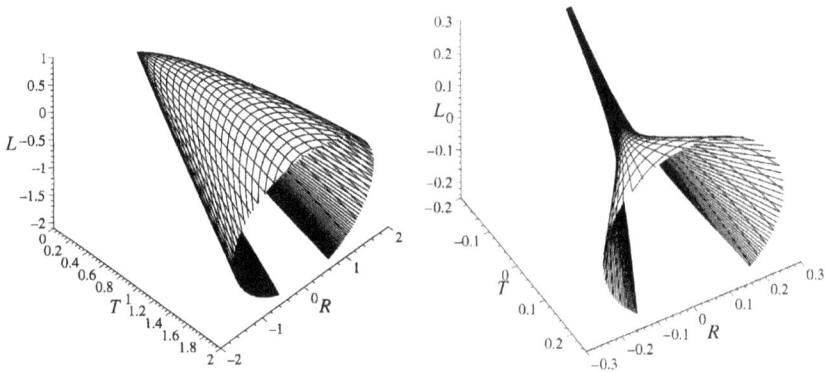

**Figure 5.6.** The shape of the universe as specified by a solution of the 4D Einstein equations of general relativity can be better appreciated by an embedding in flat 5D space. Such embeddings have to be done carefully, respecting the nature of the matter (a constant $\alpha$ determines the equation of state, where $\alpha = 3/2$ corresponds to zero pressure and $\alpha = 1/30$ corresponds to a negative pressure). In each of these two diagrams, the mesh drawn on the model universe has the ordinary time running along the growing shape, with the ordinary radius running orthogonally. Each model 4D universe is embedded in a flat 5D manifold, whose essential coordinates are $T$, $R$ and $L$ (representing alternative measures for the time, the radius and the fifth axis). The first picture shows the basic 4D Einstein–de Sitter universe, which evolves with the shape of a parabaloid. The second picture shows an inflationary 4D universe, which evolves with the shape of a trumpet. In neither case, from the 5D perspective, is there an initial singularity or big bang.

of general relativity, except that the source is now given explicitly in geometrical terms. This means that the energy-momentum tensor is expressed in terms of derivatives of the scalar potential $\Phi$ and the ordinary 4D potentials for gravity $g_{\alpha\beta}$. (For ease of study, the ordinary partial derivative can be denoted by a comma, while the covariant derivative which takes account of the curvature of spacetime can be denoted by a semicolon.) The derivation of the required expression for the matter source requires a lengthy manipulation of the five dimensions concerned (Wesson 2006). The result is

$$8\pi T_{\alpha\beta} = \frac{\Phi_{,\alpha;\beta}}{\Phi} - \frac{\varepsilon}{2\Phi^2}\left\{\frac{\Phi_{,4}g_{\alpha\beta,4}}{\Phi} - g_{\alpha\beta,44} + g^{\lambda\mu}g_{\alpha\lambda,4}g_{\beta\mu,4}\right.$$

$$\left. - \frac{g^{\mu\nu}g_{\mu\nu,4}g_{\alpha\beta,4}}{2} + \frac{g_{\alpha\beta}}{4}[g^{\mu\nu}{}_{,4}g_{\mu\nu,4} + (g^{\mu\nu}g_{\mu\nu,4})^2]\right\}. \quad (5.6)$$

This may look a little clunky, but it includes all known forms of matter plus others which are yet to be studied. Any 4D property of matter can be read off from the 5D geometry. For example, the conventional density $\rho$ (or $T_{oo}$ in the above) is basically the second time derivative of the scalar field which forms the fifth dimension, plus some other terms which depend on derivatives with respect to the fifth coordinate of the 4D gravitational potentials. Irrespective of whether (5.6) is taken to be the right-hand side of the field equations (5.5) of general relativity, the mere existence of such a relation is of considerable philosophical importance.

Einstein was right: the coarse "wood" of what we call matter can, if so desired, be transformed into the fine "marble" of geometry.

## 5.6    Conclusion

The word "matter" has extended its purview over the history of physics, and now hosts a range of properties corralled by a few equations.

In this chapter, we have looked at several aspects of matter. The equation of state, typically between the pressure and the density, is a catch-all relation which is convenient to use where microscopic information is lacking (Section 5.2). The origin of matter remains a mystery, though the evidence suggests that most of it was created (or at least reorganized) at some specific time in the past, though the nature of the 'big bang' needs analysis (Section 5.3). The fact that the pressure and density can both be either positive or negative for the 'vacuum' as measured by the cosmological constant, suggests that we should consider the possibility of negative particle mass, whose apparent absence may be telling us something important about the universe (Section 5.4). Matter may be manipulated into various forms, and if we so choose given a geometrical description in terms of an extension of general relativity from four to five dimensions (Section 5.5). This achieves Einstein's dream of unifying the gravitational field with its source, creating a monolithic mechanics.

The properties of matter, as they have evolved over time, are phenomenological in nature: terms like "density" and "pressure" are labels we have found convenient to ascribe as human investigators. They can be given a deeper rationale through the equations of physics; and as our knowledge has grown, so has the complexity and power of our equations. This is evidenced by the passage from the relations of Poisson and Laplace to the field equations of Einstein, and those of higher-dimensional versions of general relativity like space-time-matter theory. Indeed, 5D theory gives us a comprehensive description of the properties of matter using geometry as a basis. It is a complete theory of classical mechanics; though the need to incorporate the quantum attributes of particles implies that more theoretical progress can be expected.

Some philosophers, notably Russell, have sought to define matter as the stuff which obeys the equations of physics (e.g. Russell 1967). This, however, has the hydrogen-sulphide whiff of the chicken-and-egg paradox, in terms of which came first. The sensible answer to this, of course, is that they evolved together. The same applies to our understanding of the properties of matter and the equations we employ to describe them. The amount of knowledge we have about matter is actually enormous. The natural questions arise: Is there a limit to the amount of information we can usefully accrue? And will theoretical physics become so sophisticated as to become effectively self-defeating?

Already, physics as a subject is fragmented into subdisciplines, the result of the need for its practitioners to specialize in order to become passably expert. The absent-minded professor is so because his mind discards trivial things in order to focus on the more important facts of his profession. Russell was of the opinion that a limit to understanding would not emerge to impede our progress, because methods of teaching and learning become more effective with time. However, this argument appears to be flawed, because the human brain is afterall an organ of finite size. A possible way to sidestep the finite capacity of the human brain is apparent in modern science, namely the storage of information in nonbiological systems, such as the hard-drive of a computer. But while in widespread use, this ploy does not presently offer a way to address that other vital aspect of science: the creation of a theory to underlie and explain the data. It is not an accident that the great advances in science have been made by individuals using their minds. While modern computers may be useful and even indispensable in evaluating the elements of a theory, the latter still requires for its completion the inspiration and insight of the human mind. (For example, in looking for a solution to Einstein's field equations, a computer program such as GRTensor is very useful in evaluating necessary terms like the Christoffel symbols, but the use of these to winkle out an answer depends on the ingenuity of the researcher.) In discussing this problem, we are referring mainly to the scientist's conscious thought processes. However, another potential help to the advancement of science is to make better use of the researcher's subconscious thoughts. Several great scientists are on record as having solved knotty problems by some kind of subconscious ratiocination. It would be illuminating to develop a better understanding of how the human mind secretly works out its puzzles.

Of course, there is no unique and universal method by which a scientist solves a problem, even one involving purely conscious thought. Different scientists do it in different ways. It is hard to discern a common logical approach among scientists occupied with research, especially when it is of the non-mathematical variety. It is as difficult to define the "logic" of the scientist as it is to delineate the "morality" of the average person. We will return to these issues in the last chapter; but meanwhile, it is instructive to look at how the lives of some great scientists have been molded by logic or the lack of it.

## References

Barbour, J., Pfister, H. (eds.), 1995. Mach's Principle: From Newton's Bucket to Quantum Gravity. Birkhauser, Boston.

Bonnor, W.B., 1989. Gen. Rel. Grav. 21, 1143.

Campbell, J.E., 1926. A Course of Differential Geometry. Clarendon, Oxford.

Eddington, A.S., 1939. The Philosophy of Physical Science. Cambridge University Press, Cambridge.

Einstein, A., 1950. The Meaning of Relativity, 3rd edn. Princeton University Press, Princeton.

Kaluza, T., 1921. Sitz. Preuss. Akad. Wiss. 33, 966.

Klein, O., 1926. Z. Phys. 37, 895.

Mach, E., 1893. The Science of Mechanics. Open Court, La Salle (reprinted edn., 1960).

Overduin, J.M., Wesson, P.S., 2008. The Light/Dark Universe. World Scientific, Singapore.

Rindler, W., 2001. Relativity: Special, General, and Cosmological. Oxford University Press, Oxford.

Russell, B., 1967–1969. The Autobiography of Bertrand Russell (3 vols.). George Allen and Unwin, London.

Wesson, P.S., 2006. Five-Dimensional Physics: Classical and Quantum Consequences of Kaluza–Klein Cosmology. World Scientific, Singapore.

Wesson, P.S., 2008. Gen. Rel. Grav. 40, 1353.

Whittaker, E.T., 1910 (and 1953). A History of the Theories of Aether and Electricity. Nelson, London.

Will, C.M., 1993. Theory and Experiment in Gravitational Physics. Cambridge University Press, Cambridge.

# Chapter 6

# SCIENCE AND RELIGION: IMMISCIBLE?

## 6.1    Introduction

Science and religion are now widely regarded as antithetical; but the adaptability of the human mind has meant that some notable scientists have lived with — and even drawn inspiration from — a strong belief in God, so the subject warrants a short investigation.

Much of modern science was developed in western Europe, which means that its spiritual backdrop was the Judeo-Christian one espoused by the Bible. In the latter, a unique male God created the heavens and the Earth; and the female half of humankind was supposed to follow from attentions paid to a discarded rib. This quaint history is paralleled by others from different places. For example, according to the traditional beliefs of Australian aboriginals, humans were ejected from the mouth of an anguished snake; while following the ancient teachings of south-east Asia, human civilization emerged from the curdling of a bowl of milk. There is a tendency today to dismiss these ancient cosmologies as simple-minded and anthropocentric. Certainly, the idea of a big bang — from which everything emerged in the explosion of a singularity — is more logical and antiseptic. But while atheism is in the ascendancy, religion shows a stubborn refusal to disappear. At present, most of those scientists who feel a need for spirituality tend to separate their belief in God from their belief in science. As we will see, however, it was not always so.

## 6.2    Newton

In the England of the 1600s and 1700s, the church played a regular part in the lives of almost everybody. Newton (Figure 6.1) believed in God, and was prepared to say so in print. However, most of his years at Cambridge were spent in scientific thinking and experimentation, and he was the first person to give a comprehensive and mechanistic account of the world. In middle life, he appears to have been in the mental doldrums, and tinkered with alchemy. But the publication of his Principia in 1687 laid the foundations, via its laws of motion and gravity especially, for a coherent view of nature in which religion really plays no part. He defended the scientific products of his mind with what modern scholars view as an excess of egoism. It is now commonly acknowledged that he unreasonably disparaged the contributions to mechanics of Hooke, and that credit for the invention of the differential calculus should be shared with Leibnitz. In the latter part of his life, Newton (1643–1727) was in charge of the English mint, and so indirectly responsible for the money in the pockets of a populace that was largely ignorant of science. Indeed, even other natural philosophers (as scientists were called in the 1700s), were in awe of Newton. His stature is

**Figure 6.1.** Newton, father of physics, stated a belief in God.

perhaps best caught in the well-known poem of William Wordsworth (1770–1850, *The Prelude*, book iii at line 61):

Where the statue stood
Of Newton, with his prism and silent face,
The marble index of a mind for ever
Voyaging through strange seas of Thought, alone.

So while Newton may not have been a very likeable person, he at least set in the public mind the image of the unapproachable scientist which has survived into modern mythology.

## 6.3    Einstein

The epitome of scientific genius, Einstein was a spiritual man but not religious in a conventional manner. He sometimes used the word "God" in a playful sense; and his affable but slightly naive personality must have served him well in the political and sociological turmoil of the times in which he lived (Halpern 2004). He had the kind of mind which (like Feynman later) was able to cut through a confusing superstructure of comment and isolate the basic scientific issue. His book *The Meaning of Relativity* (1950) is remarkably slim. Einstein can justifiably be termed brilliant, because he was able to produce a string of fundamental results without apparent strain; and was well known for being able to relax in ways disconnected from physics, like playing the violin and sailing. The Europe in which he was living as a young man was undergoing a period of political meltdown, which affected many scientists adversely. (Planck, for example, had his livelihood and familial happiness destroyed by Nazi zealots.) Einstein's Jewish ancestry did not help, and after working in several European countries he eventually moved to the United States of America. At Princeton, he searched for a unified theory of fields and matter (see elsewhere), while also attempting to ban the development of the atomic bomb, which he had unwittingly made possible by his formulation of the famous law $E = mc^2$. (For a scientific biography see Gribbin 2005; the present discussion is short because much has been written about Einstein — and Newton — already). The noted formula gives the energy stored in a given mass, and is by conventional standards enormous because of the large value for the speed of light. His involvement in the peace movement and other political issues meant that Einstein in his later years was often regarded as being sociologically slightly inept. However, he had the wisdom to decline an invitation to become the figurehead of the state of Israel, which came into existence in 1948. Einstein, who lived from 1879 to 1955, was ill in his later years. In an irreligious but scientifically justified act, he donated his brain to biological analysis, and it is in fact still available for study. But examinations to date have not revealed anything special about its structure, implying that anybody might in principle become a genius.

## 6.4   Eddington

A contemporary of Einstein, who during his lifetime lived partly in the great man's shadow, Eddington has a quiet fame which continues to grow. In terms of spirituality, Eddington was a Quaker. This movement is arguably more philosophical than religious in nature, somewhat in the manner whereby Buddhism is more concerned with a way of life than the theology of godhead. Eddington (1882–1944) went regularly to the Friends Meeting House in Cambridge, where congregations operated (as they still do) in a democratic fashion without a head as such. This form of egalitarianism is one of the basic tenets of the Quaker way, along with the renouncement of violence as a means of settling disputes. The name is reputed to have arisen when in 1650 the movement's founder (George Fox) became embroiled in a controversy with the English authorities, whom he suggested should be more humble and quake before the

Lord. Eddington himself did not make public very much about his religious convictions or his activities as a member of the Quakers. He did, however, feel justified in refusing to enlist in the armed forces when, in the middle of his life, Britain and Germany declared war on each other. Eddington was, by the accounts of the few colleagues who knew him, a modest man.

He lived for a good part of his career with his sister in the stone house on Madingly Road west of Cambridge, which is nowadays one of the two main buildings which accommodate the astronomy department of the University. (The other is now called the Hoyle building, and is a modern structure planned without the aid of an architect by Fred Hoyle, who after Harold Jeffreys succeeded Eddington as Plumian Professor of Astronomy.) But Eddington was not averse to standing up for his scientific beliefs. He had a long-running low-level argument about the application of physics to astronomy with James Jeans. The latter was a mathematician and musician, who became well known through his assertion that God ought to share both of these interests. Nowadays, the other man is mainly remembered for the Jeans mass, which is the typical one formed when a cloud fragments under gravity to form objects like stars and planets. The original calculation of this was in fact faulty, and as Hoyle later remarked, Eddington got his calculations right whereas Jeans was prone to sloppy analysis and speculation. (He once suggested that the shapes of spiral galaxies indicated that they were the sites where matter was being poured into our universe from an extraneous dimension; which is a possibility, but was not backed up by Jeans with any significant analysis.) Eddington himself became well known by writing several popular books on cosmology (e.g. *The Expanding Universe*, reprinted in 1958). These are excellently done, and a joy to read. This was in contradiction to the public lectures on which they were based, which were reportedly marred by the introspective character and mumbling delivery of the speaker. The continuing appeal of Eddington's books — both popular and technical — is due to his profound understanding of cosmology.

Following the formulation of the general theory of relativity by Einstein, the appreciation of it was very limited in the 1930s and 1940s, due to its intensely mathematical nature. Eddington undertook the translation of the theory into English, and attempted to educate the scientific community in Britain and the United States of America about its far-reaching implications. Asked by a reporter if it were not true that only three people in the world understood the theory, Eddington responded facetiously by asking who the third person might be. This illustrates that while Eddington was a staunch supporter of Einstein, he was aware of his own notable status within the physics community. In fact, Eddington was not averse to disagreeing with Einstein: the former regarded the cosmological constant as the foundation of gravitation as applied to the large-scale universe, while the latter was at pains to try and disregard what he considered to be "the greatest blunder" of his life. Also, Eddington in his later years attempted to unify gravitation with the emergent field of quantum physics by moving away from the field-theoretic approach, to one that was numerological in basis and used simple equations. (For example, he tried to calculate the masses of the electron and proton as the roots of a quadratic equation with coefficients related to the fine-structure constant of atomic physics; see Section 3.3.) These later forays by Eddington into the grey area between general relativity and quantum theory are easy to criticise today. However, we do not know exactly what was in Eddington's mind, or how far his considerations had proceeded, because his last book *Fundamental Theory* was put together posthumously from incomplete notes found in his desk (Slater 1957; Batten 1994). Had he lived, it is conceivable that he might have developed his numerological considerations into a unified theory as self-consistent and as tight as his earlier works. As it is, *Fundamental Theory* is a flawed testament to a great thinker. It is still pored over by researchers, who hope to decipher its cryptic contents — something like a Rosetta stone for physics.

## 6.5    Milne

While Eddington was working on the fundamentals of cosmology in Cambridge, Milne (Figure 6.2) was doing the same in Oxford. However, the spiritual stances of the two men were quite different, as Milne was an avowed Christian. In contrast to Eddington's extensive and non-religious writings, Milne produced just two noteworthy volumes: *Kinematic Relativity* (1948) and *Modern Cosmology and the Christian Idea of God* (1952). In the first of these, he used an elegant method based on group theory to construct a model universe. In it, an observer on one typical galaxy estimates the distances to others by

**Figure 6.2.** Milne (1896–1950) was a devout Christian who formulated a viable relativistic model of the universe.

sending out and receiving pulses of light. Shifting the observer to another galaxy and insisting that the same kind of data be acquired defines the mechanics of an isotropic and homogeneous (uniform) universe, in which each galaxy increases its distance from another one in proportion to the age. Motion where distance is proportional to time is 'free', meaning that no forces act (the galaxies in Milne's model are supposed to have infinitesimal masses, so their gravitational interactions are negligible). This kind of motion forms the subject of kinematics, as opposed to dynamics where forces are in play. It is remarkable that Milne was able to arrive at a valid model for the universe using only an elementary application of group theory, without appeal to the field equations of gravitation as formulated by Einstein. Today, however, Milne's model is usually regarded as a kind of limiting case for the Friedmann–Robertson–Walker (FRW) class of models. It actually satisfies Einstein's equations of general relativity.

For a uniform universe described by a perfect galactic fluid, Einstein's equations reduce to two relations named after Friedmann. One gives the density $\rho$ and the other the pressure $p$, when the distance between galaxies increases with time in proportion to a scale factor or length that depends on the time, $S(t)$. Thus:

$$8\pi G\rho = \frac{3}{S^2}(kc^2 + \dot{S}^2) - \Lambda c^2 \qquad (6.1)$$

$$\frac{8\pi Gp}{c^2} = -\frac{1}{S^2}(kc^2 + \dot{S}^2 + 2S\ddot{S}) + \Lambda c^2. \qquad (6.2)$$

Here $k$ signifies the curvature of ordinary, three-dimensional space, and is normalized so that it takes on the values 0 or $\pm 1$. A 'flat' FRW model is really only so in its 3D sections. Most FRW models are curved in 4D spacetime, even if they are flat in 3D. The 4D curvature is related to the presence of ordinary matter (specified by $\rho$ and $p$) and/or the presence of a finite cosmological constant $\Lambda$. The latter is included explicitly in the above equations, but can if so desired be regarded as implicitly defining the density and pressure of the vacuum fluid, as discussed in Chapter 5 (see Figure 5.4 for plots of $S$ versus $t$). If so, its properties do not change, whereas those of conventional matter do, as dictated by the time derivative of the scale factor (denoted by an overdot, where in models with a big bang the age is measured from that event). In regard to the Milne model, an inspection of the Friedmann equations with $\Lambda = 0$ shows that they are satisfied by two simple choices of parameters: Minkowski space has $\rho = p = 0$, $k = 0$, $S$ = constant; Milne space has $\rho = p = 0$, $k = -1$, $S = t$. Neither of these solutions is particularly realistic, since both are devoid of matter. However, while the first is static (and in fact the laboratory space of special relativity), the second is expanding in a way not too dissimilar from the motions of real galaxies. It has other properties too, such as the absence of observation-limiting horizons, which are compatible with modern astrophysics. Therefore, Milne space is sometimes employed as a model for the universe in the limit where the matter in the galaxies and the cosmological constant can be neglected. However, its main impact is of a more philosophical nature: Milne space is mathematically equivalent to Minkowski space. That is, there is a change in the coordinates of distance and time which can be carried out on Minkowski space to give Milne space (see e.g. Rindler 1977 for the detailed calculation). In technical language: Minkowski space and Milne space are isometric ("equal measure") with respect to each other. In colloquial language: a static, flat laboratory space like the surface of a table and an expanding, curved space like the universe are the same thing, looked at in different ways. However, we note that while the equivalence of the two M spaces is remarkable, it is not miraculous. Other 4D isometries are known. And in 5D there is a similar case (Wesson 2008), where a big bang expanding universe with massive galaxies is isometric to a perennial, static and empty one.

For cosmologists with religious convictions, it might be tempting to draw a parallel between such a scientific situation and the corresponding theological one, involving the biblical version of the creation of the world and the buddhistic view of the permanent void. But in practice, such speculations lead to nothing of value for either science or religion. In Milne's case, his scientific book showed a valid world-model based on logic; while his religious one added little of scientific value, and may even have caused some dissention among other believers with different convictions. After all, the Milne universe starts in a big bang, which by one application of the rules of religious logic might be interpreted as the death of God.

The inference from our brief survey of Newton, Einstein, Eddington and Milne is that the minds of gifted people work independently where it comes to science and their nonscientific beliefs. There is little evidence that the technical achievements of these people have been furthered or enriched by their sociological attitudes or their religious beliefs. It should also be stated that the four people we have considered were chosen not only because of their scientific prowess, but because of the attention paid in history to their nonscientific activities. It would be easy to write down a list, with double the number of names, of individuals who had no professed views on spirituality or religion. This does not mean that other well-known scientists are calculational machines unmoved by the beauty or ugliness of the world in which they live. The typical ageing cosmologist wonders just as much about death as anybody else; and the prospects of his elements one day returning to the interstellar space from which they originated is not something which offers immediate comfort (see Wesson 2002 for an account of gallows humor). The main spiritual advantage of working on cosmology — as opposed for example to earning a living on a car assembly line — is that one occupation is more interesting than the other. The urge to avoid boredom and broaden the mind is probably the biggest unappreciated factor in the development of cosmology (see Leslie 2001 for an account of human thinking about the universe). We choose to complement the preceding survey of how spirituality may exist side-by-side with physics, by turning to a person who is widely regarded as the father of atheistic science.

## 6.6    Laplace

This enormously gifted French mathematician (Figure 6.3) followed the English physicist Newton, and formulated many of the equations which underlie modern science. He enjoyed considerable sway in putting forth his views on physics to a largely uneducated public, and several parts of his treatise on the analytical theory of

**Figure 6.3.** Laplace (1749–1827) gave exquisite mathematical form to much of physics, and had "no need" for God.

probability (1812) received popular attention. Among Laplace's writings, we find the following remarkable statement:

> Given for one instance an intelligence which could comprehend all the forces by which nature is animated and the respective positions of the beings which compose it, if moreover this intelligence

were vast enough to submit these data to analysis, it would embrace in the same formula both the movements of the largest bodies in the universe and those of the lightest atom; to it nothing would be uncertain, and the future as the past would be present to its eyes.

This is an early but convincing presentation of the scientific philosophy which came to be known as determinism. In a Newtonian world, where the laws are specified precisely, there is a chain of events that links the microscopic to the macroscopic, and leads to the conclusion that people are also governed by determinism. Indeed, the whole cosmos and its inhabitants form a gigantic clockwork — intricate perhaps, but mechanistic. Of course, many people object to this, saying that they can choose what acts they do or do not perform, and that they have free will.

The question of determinism versus free will has been debated for centuries, and this is not the appropriate place to give a detailed account of it. But some of the objections to determinism are specious, and need to be briefly debunked. For example, a common view today is that quantum mechanics somehow provides a loophole, through which we can wriggle out of the straightjacket of determinism and enjoy free will (see the books by Deutsch 1997 and Bell 2004). This is probably incorrect. For even if the logic of quantum mechanics proves to be different to that of classical mechanics, it is still a form of logic, and thereby provides a linkage between events. And it does not matter if we label such events as cause and effect, since all that is needed is the existence of a certain relationship between events in order to establish determinism. A related argument that is sometimes used to defeat determinism is that Heisenberg's uncertainty relation implies an inherent level of indeterminacy at the microscopic level. However, certain physicists have always found this view abhorrent. Einstein summed up the situation by declaring that in his view "God does not play dice with the universe." Some physicists have politely disregarded this view as old-fashioned. But recently, it has been shown that a Heisenberg-type relationship between the mechanical parameters of four-dimensional spacetime can be understood as the 'left-over' bits of completely deterministic laws in a five-dimensional world, of the type needed to unify gravity with the interactions of particles (Wesson 2004). These comments mean that it is misleading to dismiss determinism by a glib appeal to quantum mechanics. After all, if free will is really the way of the world, it ought to be possible to show it by scientific reasoning. In fact, most people who disregard determinism do so not for scientific reasons at all, but because they have an instinctive belief in free will. It is instinctive for some people in the same manner as many other automatic responses of the human brain — a belief of the same kind as the one involving God.

There is, though, a valid question which an adherent of free will can ask of a Laplacian determinist. It is: "If the world is deterministic, why is it that I cannot foresee the events of the future?" This qualifies as a good, scientific question. Related ones are: "Why do I have only an incomplete recollection of events in the past?" and "Why when I do scientific research do I feel that I am discovering new information?" The answers to these questions are actually to be found in a consideration of the views of Laplace (1812) quoted above. There is in fact one answer to all criticism of determinism of the preceding type, and it is simple and straightforward: "I am not Laplace's super-being, and my human brain is of limited power and has imperfect abilities." This answer reinstates determinism (though perhaps in the context of Plato's perfect ideas), and is in line with other imperfect operations of the human mind of which we are all distressingly aware. The main objections to the opinion just outlined are nonscientific ones, rooted in the egoism of some researchers, and the belief of others that they are made in the image of a perfect God. For the record, this writer and his colleagues have enough humility (at least most of the time) to admit that their minds are not perfect, and that they just might be pawns in a darkly-seen deterministic world.

Returning to Laplace, the advocate and defender of determinism, we can reflect on what he said about religion. His views on this are shared by a large number of modern scientists. And when cosmologists versed in general relativity try to communicate with the heads of organized religion such as the Pope, the result is usually a philosophical disconnection (see Figure 6.4). When Laplace presented his conclusions about the natural world to the Emperor of France, Napoleon Bonaparte, that individual asked Laplace where God fitted into the picture. To which the scientist replied: "I have no need of that hypothesis."

**Figure 6.4.** Pope John-Paul II chatting in 1985 with an international group of cosmologists. Most of the latter look bemused (including the author at the extreme right), because their suggestion that the world began in a big bang had been met with the opinion that a better theory could be found in the Bible. This is an example of the immiscibility of religion and science.

## 6.7    Conclusion

The temerity of Laplace is objectionable to some people but admirable to others, depending on where they are located in the spectrum of religious beliefs. Laplace's words hang like a judgment over past and present scientists. They say, in effect, that Newton was mistaken in his advocacy of God, that Einstein sat on the fence, that Eddington was confused, and that Milne was completely misguided.

For those scientists who hold religious beliefs, they can decide for themselves how far Laplace's barbs penetrate their sphere of spirituality.

Presently in the western world, there is a tendency to regard religion as obsolete. Many people appear to believe that there is a practical if amorphous philosophy of the world, formed by an amalgam of objectivity, logic and physics. This nondescript philosophy grows steadily with time, via the scientists, engineers and teachers who practice it — somewhat like a cultural snowball that grows in size as it rolls onward through history. This kind of mechanistic view of the world may, in a formal sense, be "correct". But it is discouragingly drab. It lacks the pizzazz of other fields of human endeavor; and does not excite the positive feelings that attend (say) viewing a beautiful painting, watching an intricate ballet, or listening to a deep symphony. This is puzzling. Writers interested in the sociological and historical aspects of science opine that its practitioners are passionate people who realize that they are involved in a cultural exercise (Robinson 2009, Shapin 2009). And many scientists avow that doing research is among the most fascinating things they know.

Is it possible that in our preoccupation with the materialistic benefits of science we have misjudged its essential character? Is it plausible that the development of a physical theory is not merely a plodding application of algebra, but instead akin to the creation of a fine painting, the choreography of a new dance, or the composing of a fresh piece of music? In other words, does the essence of Science lie closer to what we usually call Art?

## References

Batten, A., 1994. Quart. J. Roy. Astr. Soc. <u>35</u>, 249.

Bell, J.S., 2004. Speakable and Unspeakable in Quantum Mechanics, 2nd edn. Cambridge University Press, Cambridge.

Deutsch, D., 1997. The Fabric of Reality. Penguin, London.

Eddington, A.E., 1958. The Expanding Universe, reprinted edition. University of Michigan Press, Ann Arbor. [For a bibliography of Eddington's books, see Wesson, P.S., 2000. Observatory <u>120</u>, 59.]

Einstein, A., 1950. The Meaning of Relativity, 3rd edn. Princeton University Press, Princeton.

Gribbin, J. and M., 2005. Albert Einstein and the Theory of Relativity. Chamberlain Bros., London.

Halpern, P., 2004. The Great Beyond: Higher Dimensions, Parallel Universes, and the Extraordinary Search for a Theory of Everything. Wiley, Hoboken, N.J., p. 171.

Laplace, P.S., 1812. Analytical Theory of Probability. Courcier, Paris.

Leslie, J.D., 2001. Infinite Minds: A Philosophical Cosmology. Clarendon, Oxford.

Milne, E.A., 1948. Kinematic Relativity. Clarendon, Oxford.

Milne, E.A., 1952. Modern Cosmology and the Christian Idea of God. Clarendon, Oxford.

Newton, I., 1687 (translated by A. Motte from Latin into English, 1729). Philosphiae Naturalis Principia Mathematica. Societatus Regiae, London.

Rindler, W., 1977. Essential Relativity, 2nd edn. Springer, New York, p. 205.

Robinson, K., 2009. The Element: How Finding Your Passion Changes Everything. Penguin/Viking Press, New York.

Shapin, S., 2009. Science as a Vocation. University Chicago Press, Chicago.

Slater, N.B., 1957. The Development and Meaning of Eddington's 'Fundamental Theory', Including a Compilation from Eddington's Unpublished Manuscripts. Cambridge University Press, Cambridge.

Wesson, P.S., 2002. The Interstellar Undertakers. Vantage, New York.

Wesson, P.S., 2004. Gen. Rel. Grav. 32, 451. [See also ibid., 2006, 38, 937.]

Wesson, P.S., 2008. Int. J. Mod. Phys. D 17, 635.

## Chapter 7

# WEAVING THE WEFT

The question we are now in a position to pose, and hopefully answer, is both profound and provocative: Is science discovered or invented?

By this, we do not mean to imply that science is a dream or fantasy. Rather, we are following the path of the eminent astronomer Sir Arthur Eddington in the 1930s, and asking if the biological and psychological aspects which necessarily attach to us as humans introduce a subjective element into what we usually regard as objective research. It is difficult to answer this in a completely quantitative fashion, and attach a value between 0 and 100 percent. But an increasing number of people are asking this question, as our theories of subjects like quantum theory and cosmology become ever more abstract, taking on the appearance of mind weaving.

When a weaver at the loom starts to create a garment, the foundation lines are laid out in the form of the warp (Chapter 1). Then the weft is added at right angles, carrying with it the colour, texture and other properties that give the product its characteristic appearance. The scientist is in some ways in a similar situation: the basic laws of physics are there, but on these is often built a complicated fabric which owes more to interpretation than anything else. However, interpretation is idiosyncratic, and this is why certain subjects are discussed in widely different versions. As an example, most cosmologists learn and believe the laws of gravitation as formulated in Einstein's theory of general relativity; but some accept the origin of the universe in a big bang, whereas others prefer to replace it with a non-singular event. If cosmology consisted only of the application of cast-iron laws to the natural world, this freedom of interpretation would not exist. And in addition to the flexibility of interpretation, there is also the possibility that the underlying laws may themselves be open to reformulation. It is reasonable to ask about the apparent sureness of science; and to inquire if in fact it is more fluid than commonly assumed — more like other fields of human endeavour which we classify not as Science but as Art.

Many scientists believe that in doing research they are discovering new facts about nature, in the same manner as a prospector tracks through the wilderness and occasionally turns up a nugget of gold. It is easy to see why this attitude prevails: if a scientist — perhaps a cosmologist or quantum theorist — manipulates his equations and finds some neat result which he believes is not known to other scientists, he is likely to say that he has "discovered" the "new" result. But the terminology here is flawed even at a surface level. For the scientist is probably using the same equations available to all of his fellows — for example those of general relativity or those describing the interactions of particles — so his "discovery" is not a fundamental breakthrough, but merely the uncovering of a result that was already inherent to his theory, and "new" only in the sense that he has realized its significance before his colleagues. In addition, the theory used by the average scientist does not have some holy, magical or otherwise mysterious origin, but was produced in the mind of one of his (usually more clever but long-dead) compatriots. For example, Einstein struggled for some time in formulating the relations that underlie what we now call general relativity. Indeed, his first attempt at this was unsatisfactory; and it was only after considerable mental gymnastics that he proposed what eventually became known as Einstein's field equations. Those equations are not God-given. If they are hallowed in any sense, it is by time and widespread acceptance. But we should not forget that they were hatched in the mind of a man — admittedly an unusually perspicacious one — but a man for all that.

Eddington, who lived at the time when Einstein was thinking about relativity, understood better than most of today's workers that science is largely a product of the human mind. Eddington actually started his career at Cambridge working on observational data, and later headed one of the expeditions to observe the eclipse of the Sun which validated Einstein's then new theory of general relativity. He cannot therefore be accused of being a prima donna of theoretical physics, unconcerned with reality. However, as he aged, his views changed. He realized that even an observation or experiment necessarily involves a certain theoretical framework for its interpretation. And as he considered things in more and more detail, he became convinced that much of science owes its content not so much to the external world as to the internal machinations of the human mind. In the latter years of his life, he was preoccupied with the idea that science might be an intellectual exercise, and that its content might be attributable to mathematics and the power of the brain. It was the invented aspect of science he was considering when he expressed the metaphorical statements for which he is now largely remembered: that the scientist who finds a sign on the beach of knowledge recognizes it as his own; and that the stuff of the world is mind-stuff.

Reluctant as many people are to admit that science is subjective, it becomes increasingly easy once the first step is taken. This author was initially reluctant to accept Eddington's views; but found them progressively more reasonable when he reminded himself that Man is but a species of animal that necessarily processes sense data through the biological equipment he has inherited by millions of years of evolution, as recognized by Darwin (Figure 7.1). Our view of the external world is filtered by our senses; and the internal picture we create as a model of the world depends on the ingenuity of our minds. The minds of some individuals are more fertile than others, and it is to those of this kind with an aptitude for mathematics that we owe much of modern physics. Such people are driven to extend the fabric of physics and to strive for novel intellectual heights (Figure 7.2). That is why we have, currently, several alternatives to Einstein's theory of general relativity that go beyond that subject. The inventiveness of the human mind also finds expression in more concrete forms, such as new and very intricate experiments. We will argue below that these serve mainly as complements to, rather than substitutes for, the purely intellectual results of research.

**Figure 7.1.** Darwin lived to see the widespread acceptance of his theory of the origin of species by differentiation, though his proposed mechanism by natural selection of the fittest was more controversial.

Physics is the central subject of the 'hard' sciences, and it owes this stature largely to its use of mathematics. Several workers have argued that mathematics is essentially a kind of language. This may be disputable, but it has a certain level of validity and is certainly a practical definition. (At the International House of the University of California at Berkeley, the periodic dinners for visiting scholars are

**Figure 7.2.** Hoyle was a Cambridge astronomer who followed in the steps of Eddington, Jeffreys and Dirac. This mosaic shows him scaling the structures of conventional thinking, reaching for a higher understanding of the universe. (It is part of a piece called *Modern Virtues* by Boris Anrep, which was completed in 1952 and is located in London's National Gallery.)

served at tables which are each identified by a label indicating the language, and one is denoted "Mathematics".) Even though the nature of mathematics is not completely understood, we can in a more general sense call it a code — a code by which we attempt to increase our understanding of the world. Given this rather loose definition, we can begin to see that its attendant subject — physics — is closer to the arts than is widely appreciated. Certainly some physicists dream about equations, in the same way that some artists dream about their canvases. In fact, if we regard the arts and the sciences as both relating to our view of the world, we can discern a kind of continuum.

To see this, let us for the moment regard a painting as a kind of code in colour and form, which is a representation of the world. (Apologies might be in order for calling the roof of the Sistine Chapel a code; but on a very general level this is true of fine art, and especially of abstract art.) Then we can regard a painting — by virtue of its direct use of shapes and colour — as forming one end of a kind of spectrum of codes. Within the category, we can argue that a child's stick-man is the simplest form, while renowned works of art that incorporate subtle symbolism are a more complicated form (Figure 7.3). Progressing along the spectrum of codes to its more abstract parts, we come to prose and poetry, which are necessarily written in some colloquial language. If the language concerned is not that of the person viewing the text, it is almost obvious that writing in whatever form is a code (Figure 7.4). Again moving along our spectrum, we can argue that dance choreography is also a code, which while it uses stylized human figures is only readily interpretable to those who have studied ballet (Figure 7.5). Dance is often set to music, and in the latter we have an unmistakable code. Whether it is the simple riff of the guitar in a pop song, or the complicated harmony of the instruments in an orchestra, the notation for music is a distinct

**Figure 7.3.** In 1839, J.W.M. Turner painted *The Fighting Temeraire*. Like many works of fine art, it is not only a representation of what is seen but carries a message. In this case, a steam tug tows the sailing vessel *Temeraire* (a veteran of the battle of Trafalgar) to the wrecker's yard, symbolizing the changes caused by the industrial revolution in England.

code (Figure 7.6). While we can translate the notes on a stave into sounds, we cannot hear the interaction of the atoms that make up a chemical compound. However, the nature and configuration of the microscopic ingredients of matter are responsible for its chemical properties, and science has developed its own way of encoding this information (Figure 7.7). Chemistry represents in some ways the transition from the arts to the sciences. Interestingly, its ancient form of alchemy was regarded more as an art form, and nowadays stands in relation to chemistry as astrology does to astrophysics. Modern chemistry is certainly scientific, as evidenced by the public's faith in pharmaceuticals. Returning to our spectrum, we now move

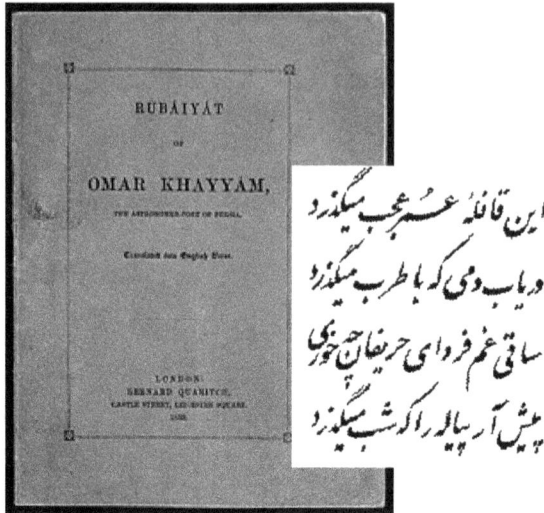

**Figure 7.4.** Poetry, like prose, is a kind of written code for human experience. The original Persian text for the Rubaiyat of Omar Khayvan is as unintelligible to the average western reader as the tensor calculus of the theoretical physicist.

**Figure 7.5.** The choreography for the dance La Cachucha, by Friedrich Albert Zorn. Accompanying the music are stylized human figures. These are in essence a code for the dance, which can be compared to the mathematical expressions used in physics to describe the behaviour of matter.

**Figure 7.6.** Page 13 of the score of Beethoven's ninth symphony. When performed, this piece of music is familiar to many people, and is used as an anthem in Europe. However, as a musical code it is analogous to the equations of the physicist.

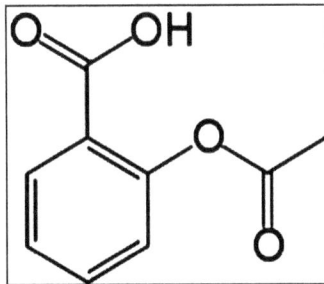

**Figure 7.7.** A representation of the drug aspirin (2-acetoxybenzoic acid). It is essentially a code, written in the language of chemistry, and as such is similar to the mathematical relations used by physicists to describe the mechanical properties of the world.

naturally to the enormous subject of physics, the doyen of the sciences. As noted, its characteristic code is mathematics, which has to be regarded as the most successful invention of its kind (Figure 7.8). It is certainly more universal than any of the colloquial languages used on the Earth, including English. However, a lay person looking at a page of tensor calculus might well regard it in the same way as a page of Persian prose (compare Figures 7.4 and 7.8). This somewhat proves the point, namely that there is a continuum or spectrum of subjects involved here, which runs smoothly from the arts to the sciences.

Our spectrum as outlined above could be extended. At the technical end, we could add the burgeoning field of cryptography, whose aim is to encode information in a way that makes it undecipherable to the uninitiated. At the pictorial end, we could add the images scratched into the rocky walls of the homes of our ancient ancestors, which predate writing and whose purpose is largely unintelligible to us. However, enough has been said to show that there is a commonality in all of the creative endeavours of mankind. Incidentally, those who practise both the arts and the sciences report that the feeling which attends the creative act is similar no matter how it is applied. (The cliche holds that creativity involves 10% inspiration and 90% perspiration.) In fact, we have now completed the first stage of the thesis, that science can be regarded as a creative thing akin to fine art.

The second stage of the thesis will involve a more detailed look at that most basic of sciences — physics — to see what makes it 'tick'. To be precise, we wish to marshal certain insights about the way in which physics is done, which show that it is essentially a creation of

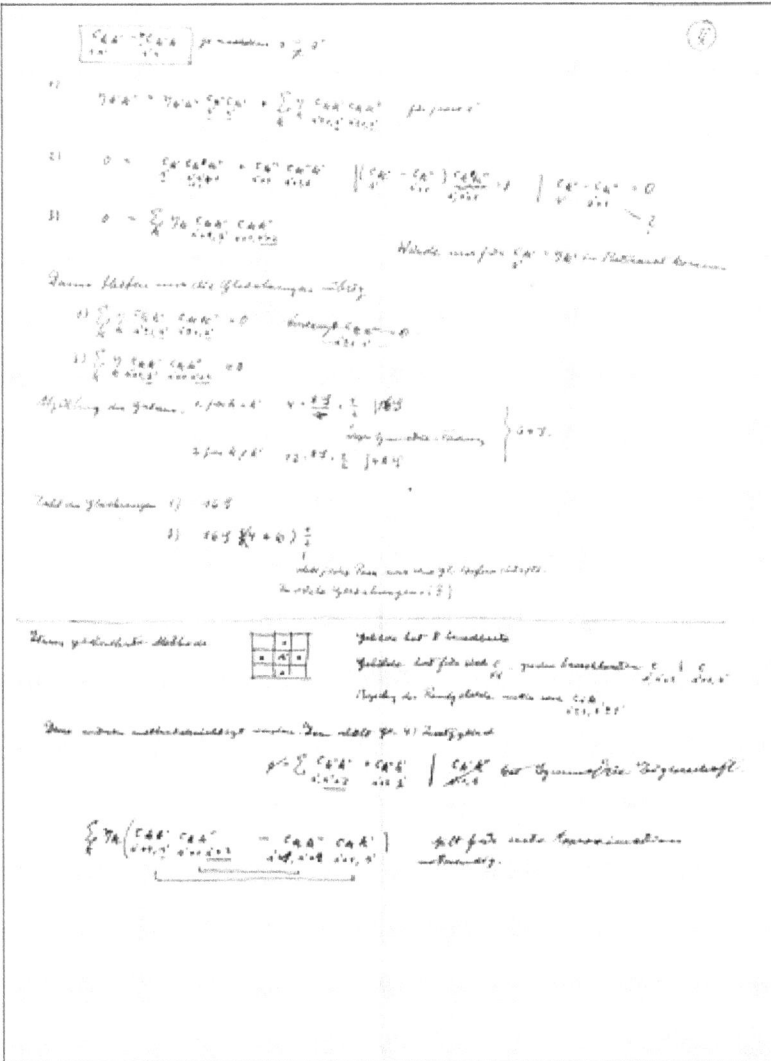

**Figure 7.8.** A page of calculations by Einstein on his theory of relativity. The equations involve tensors, and the box diagram is a pictorial representation of one such. Mathematics is the language of physics; the equality sign is its most important element of syntax; and a tensor equation guarantees that different methods of measurement do not affect the resulting description of nature.

the human mind. Since this thesis runs counter to the widespread belief that it exists independent of us, and that all we do is discover it, we will have to look at physics with new eyes. But, that is what the scientist is supposed to do anyway, is it not?

In order to make progress, it is necessary to take a break from philosophy and recall some physical facts. A judgement about the relationship between philosophy and physics will be made shortly. But for our judgement to have meaning, certain old data need to be looked at in a new light.

Puzzles in science, of the kind we examined in Chapter 2, tell us a lot about the subject, in the same way as a physician learns about the human body when it succumbs to illness. Certainly, paradoxes like the one promulgated by Olbers with regard to the darkness of the night sky, are blights on the logical progression of physics. If astronomy in Olbers' time had been a perfect subject, he would have been able to work out the low level of background radiation due to stars in the universe. It would have been a straightforward calculation, accurate to the limits set by those parameters that depended on observations. Instead, the problem as formulated was faulty at the outset, because the importance of the finite age of the shining sources was not appreciated. The ignorance of the age factor led to totally wrong estimates, with the prediction that the night sky should be ablaze with light. This paradox lay scabrous on the body of astrophysics from at least the 1860s to the 1980s. The fact that the problem is now well understood leads us to suspect that the newer puzzles of physics may find resolutions in similar, new avenues of thought. That is, so-called paradoxes in science lie not in the external world, but in the restricted thought patterns of those who study them.

In Chapter 3, we considered the vintage and still-evolving subject of dimensions. The width, height and depth of ordinary three-dimensional space are obviously biological in origin, concepts we are almost obliged to use by virtue of human physiology, particularly the eye. Time is similar in nature, but more subtle and to a certain extent internal, so we deferred a detailed discussion of it to Chapter 4.

The primitive labels $xyz$ of 3D space were formalized long ago into Cartesian coordinates, but as labels for what humans perceive as "space" they are not unique. For example, the rectangular straight-line elements $dx$, $dy$, $dz$ can be replaced by the elements $dr$, $d\theta$, $d\phi$ of the radius and orthogonal angles of a spherical system. This is more convenient for natural systems, such as the Earth or the Sun, whose shapes tend to the spherical by consequence of the action of gravity. Further investigation shows that $xyz$ may be replaced by any well-founded system of coordinates. This lays the foundation for the Covariance Principle, which asserts the admissibility of any frame of coordinates, and by its attendant apparatus of tensors, led Einstein to the law of gravity we now know as general relativity (see Section 5.2). This theory actually incorporates time as a fourth dimension (though for algebraic reasons it is frequently numbered the 'zeroth' component), on the same footing as the length measures of 3D space. The speed of light formally allows this via the definition $x^0 \equiv ct$. The importance of spacetime lies not so much in this algebraic trick as in the realization — amply displayed by general relativity and similar theories — that the manifold we construct as a basis for describing physical phenomena is deformable and (to a degree) arbitrary. This agrees with our intuition, that the labels we choose should not affect the validity of the phenomena we wish to explain. There is, of course, a glossed-over division here between the book-keeping propensity of the human mind (which we regard as internal) and the application of this to other phenomena (which we regard as external). Leaving this aside for now, we arrive at a way of doing science which is governed by certain mathematical rules but is based fundamentally on the physical concept of dimensions. Indeed, all of the field theories of modern physics — whether dealing with long-range forces like gravity or the short-range interactions of particles — employ as a basis the labels $x^{\alpha}$ ($\alpha = 0,123$) of the physical dimensions of 3D space and 1D time. However, the structure of these field theories shows that there is nothing sacred about the dimensionality $N = 4$. Einstein's field equations of general relativity, as an example, can be considered in any number of dimensions (though they show algebraic pathologies for $N = 2$ and 3). And Campbell's embedding theorem of differential geometry shows how to go between manifolds (imaginary spaces for $N > 4$) whose dimensionalities differ by one. Hence the interest especially in 5D theories of relativity that have rich physics as a consequence of using as a foundation the Ricci tensor $R_{AB}$ ($A, B = 0,123,4$ for time, space and the extra dimension). It should be noted, though, that dimensions are not the only way to construct theories of physics. Thus the properties of particles can be described by the parameters of symmetry groups, which need not necessarily be identified with physical dimensions. Also, many aspects of the world can be described using probability and statistics, which are based on pure numbers and not physical dimensions. The conclusion we reach, from all considerations, is that the *xyz* measures of 3D space are primitive examples of a more general class; and that the dimensions of physics are essentially inventions, that we have come to adopt because of their exceptional utility.

Time as a concept has a long and prickly history in philosophy and physics. But viewed as a special type of dimension, we can use what we have learned to cut through the confusion that has surrounded it and arrive at the inference drawn in Chapter 4, namely that time is a subjective ordering device. We mean by this that it is a construct of the human mind, used to organize and thereby understand what would otherwise be a deluge of sense data. This idea has been brought forward independently by a number of people, scientific and otherwise. That some kind of ordering device is needed in science can be appreciated by considering a simple application of relativity to astronomy. That subject's vast amounts of data are acquired mainly through the passive receipt of light quanta or photons. But according to even special relativity, the path of a light ray through spacetime is given in terms of the interval or 'proper' time by the equation $ds^2 = c^2 dt^2 - (dx^2 + dy^2 + dz^2) = 0$. This means that our separate subjective perceptions of elements of time $(dt)$ and space $(dx, dy, dz)$ combine to form a more objective element of 4D separation $(ds)$ which is *zero*. The fact that the noted equation tells us that the speed of light along (say) the *x*-axis is $c = dx / dt$ is merely incidental.

The important thing is that $ds^2 = 0$ defines simultaneity, with the implication that some events are in causal contact via photons ($ds^2 \geq 0$) whereas others are not ($ds^2 < 0$). For all of the events 'connected' by photons, the algebraic condition $ds^2 = 0$ means that they are coincident in spacetime. From the human perspective, a world in which everything is happening here and now is mind-bogglingly difficult to comprehend. It is easier for us to separate things, for example along the $x$-axis in space, and likewise along the $t$-axis. Indeed, we do this automatically, and are hardly aware of it. As Hoyle remarked in relation to time, if we did not organize things temporally, it would be like listening to a Beethoven piano sonata where the score was performed by pressing down all the keys at the same instant. Hardly music. The idea of ordering can be applied to any number of physical dimensions. The experience of a person's 4D spacetime life can be formulated in terms of the propagation of a hypersurface in a 5 (or higher) D manifold. For those so inclined, this allows for an interpretation where the corporeal and spiritual segments of existence are separated by a phase change, which can if desired be identified with what is commonly called death. However, the main result of our considerations is that the concept of time is an invention of the human mind, which enables us to order an otherwise bewildering flood of impressions.

The nature of matter has been catalogued by physicists in what may seem to be a daunting degree of mathematical detail. We outlined the foundational aspects of these properties in Chapter 5. Later in the present chapter, we will pluck out an important case, and examine it using as non-technical terms as possible, in order to validate the argument that even matter is at base an invention of the human intellect. For now, we recall that the symbols we use for quantities like the density ($\rho$) and pressure ($p$) of a fluid have origins which are obscured by the mists of history. The same applies to the labels we use for what we perceive to be the properties of discrete objects, such as the mass ($m$). However, the practicality of properties of matter such as these is shown by the fact that they are still in use after hundreds of years of development; albeit that the modern usage of symbols like $\rho$, $p$ and $m$ goes far beyond that of the originals. This is evident from the current understanding of the word "vacuum". Instead of meaning the absence of matter (as in $\rho = p = 0$), the vacuum is now understood to refer to a kind of invisible fluid with rather strange properties. In Einstein's theory of general relativity, it is a fluid whose effective density and pressure sum to zero in a manner that ensures the stability of more conventional matter. In terms of the cosmological constant, the gravitational constant and the speed of light, the vacuum has $\rho_v = \Lambda c^2 / 8\pi G$ and $p_v = -\Lambda c^4 / 8\pi G$. (We here take the cosmological constant to have the physical dimensions of an inverse length squared, so the associated scale is of the order of the size of the visible universe, namely $10^{28}$ cm.) However, there could be local departures from the global equation of state $p_v + \rho_v c^2 = 0$. Then there can be creation (or destruction) of matter. Such a process about 13 billion years ago might explain the origin of the matter we observe in the present universe, though the composition of the latter is still not fully understood.

We appear to be living in a cosmic mix where about 74% is vacuum stuff, most of the rest is unidentified exotic dark matter, and there is only a sprinkling of the ordinary material we see in stars and galaxies. If this recipe is confirmed by further study, it will represent a major shift in the philosophy of science, though in a rather odd sense. For after Copernicus, our non-privileged position in space was formalized as the Cosmological Principle. This is a posh way of saying that the cosmological models we take as standard are based on the assumptions $\rho$ = constant and $p$ = constant in Einstein's equations of general relativity. These assumptions automatically rule out a centre or a boundary. (See Chapter 1. There is also nothing privileged about our place in the history of the universe, though the formalization of this via the so-called Perfect Cosmological Principle of the steady-state theory does not apply to a big bang model.) However, while it is now accepted that our location in spacetime is unremarkable, the same cannot be said for our physical/chemical location in vacuum-dominated universe models of the kind currently in vogue. This simply because as humans on a planet near a star in a galaxy, our existence is based on us being associated with ordinary matter, which only makes up about 1% of the total. In fact, for all cosmologies, a little thought shows that what we understand by being "human" is closely bound up with what we understand by the behaviour of "matter". Furthermore, as we saw before and will return to below, what the physicist understands by the word "matter" is the result of hundreds of years of directed thought by a large community of like-minded individuals. It is in some ways not surprising that the properties of matter owe more to the ingenuity of the human mind than anything else.

The human mind, however, also has a capacity for sustained, non-scientific activity of the kind associated with religion. Simplistically, science and religion can both be viewed as attempts to rationalize the world, though the emphasis in the former is on material things while the emphasis in the latter is on spiritual things. They meet, of course, in the welfare of the individual. But as we saw in our brief review of Chapter 6, there is little evidence that the subjects augment each other in a significant manner. It is possible that some workers find godhead to be an inspiration for doing science, but that kind of indirect linkage is also found in other areas of human endeavour, such as classical music. History shows that great science can be done both by those who profess a belief in God and by those who disavow his existence (e.g. Newton and Laplace). This is similar to the manner in which great music can be composed by those who follow deeply religious lives and by those who reside in non-believing communist states (e.g. Bruckner and Shostakovich). Nowadays, most kinds of science — and certainly physics — are done in a spiritual vacuum. This is largely because religion as an institution has decayed over time, due in considerable measure to the tremendous advances that have taken place in astrophysics and biophysics. The former subject puts human existence into a humbling perspective, while the latter reveals the mechanistic workings of the human body. While our account has focussed on the achievements of prominent physical scientists such as Einstein, we should not overlook the equally profound influence of life-scientists such as Darwin. The result of input from both sides is that the landscape of science today is one of calm, dispassionate logic.

There are, it is true, a few aggressive champions on the fronts of both religion and atheism. For example, the cosmologist F.J. Tipler (2007) has argued that there is a solution of general relativity which is consistent with the literal age for the formation of the Earth as given in the Bible. On the other hand, the biologist R. Dawkins (2006) has argued that atheism is the only sensible attitude, and has sanctioned the slogan "There's probably no God, so stop worrying and enjoy your life." Here, we are arguing the thesis that science is more akin than widely appreciated to other areas of human endeavour that involve invention, like classical music. By the same token, science is in some ways parallel to religion, and we can learn something from the comparison. However, while science and religion can exist side by side in some circumstances, they are basically like oil and water, and do not mix.

With the above information, we are now in a position to inquire in more detail into the central issue: Is science discovered or (at least in part) invented?

In what follows, we will take up some of the preceding topics again, but now we will add a dose of opinion to the data. The opinions expressed below stem largely from the writer's experience, but are shared by many scientists. Some are controversial, and the reader is encouraged to take issue with those with which he or she does not agree. The goal is to have a good, no-holds-barred discussion about the nature of science.

To lead off our discussion, it is instructive to recall what Einstein said in an interview with the Saturday Evening Post on 26 October 1929: "Imagination is more important than knowledge. For knowledge is limited, whereas imagination embraces the entire world, stimulating progress, giving birth to evolution." In other words, the paramount thing is the power of the human mind. Certainly, Olbers in his ruminations over the darkness of the night sky, would have made more progress with the puzzle if he had been possessed of more imagination. This is shown by the fact that J.-P. Loys de Cheseaux and Edgar Allan Poe were able to come close to a resolution of the paradox by letting their minds explore a wider range of possibilities. Modern puzzles of physics may be more technical, but are essentially similar. In particular, the problems posed by the cosmological 'constant' and vacuum fields resemble a cosmic jigsaw puzzle, where the addition of one piece would make the picture clear. Contact with aliens might be able to supply the missing piece; but we should not hold our communal breath waiting for such a hypothetical event, because the Fermi–Hart paradox shows that there is something lacking in our mental assessment of the likelihood of contact with an intelligent extraterrestrial civilization, whether located on a nearby star or in a remote galaxy. The puzzles of science serve to focus the light of reason on the faults in our own minds. Or as William Shakespeare put it in his play *Julius Caesar*: "The fault, dear Brutus, in not in our stars, but in ourselves…"

Dimensions, as they relate to the ordinary space of our perceptions, are clearly subjective. It is true that the estimation of length, breadth and height have been honed by science to imposing fineness. But the concept is primitive. A horse can accurately estimate the distance ahead when it jumps a fence; and we acknowledge that it does this by receiving sense data — primarily through its eyes — and using its brain to integrate those data in such a way that the animal's muscles lead to a sure-footed landing. This learned behaviour becomes instinctive, both for the horse and for a man.

It is, however, in a way unfortunate that much of modern science relies so heavily on the primitive concept of length. Imagine, for a moment, an alternative creature to ones like ourselves, which lacks our animalistic senses. We can visualize it as a brain in a box. Would such a creature be able to develop science? A little thought will show that the answer is Yes — provided it can access data in some form, and provided it can reason in some manner (though by hypothesis not in the same way as humans). A second question then follows: Will the science of our brain-in-the-box be the same as human science? This is a much more difficult question to answer than the first. A conjectured answer is No — because even in human science, we have cases where physical descriptions of a system are based on entirely different precepts and have almost nothing in common. A practical case concerns the distribution of galaxies in deep space, which can be analysed either using the length-based formalism of relativity, or the numbers-based formalism of statistics. It is a point in favour of the human brain that we have developed both approaches (astronomy journals are full of data using one or the other mode of description), and that cosmologists constantly struggle to relate the one system to the other. This case is a comparatively mild one of the disjunction between different ways of doing science.

A more severe case might arise if and when we make contact with an intelligent, alien civilization. It is perfectly conceivable that their way of doing science might be fundamentally different from ours. Even though they inhabit the same planet, the activities of humans and ants are fundamentally *not* alike. Many people have a reasonable expectation that S.E.T.I. (the search for extraterrestrial intelligence) will one day yield positive results. But even if this should be so, what if it turns out that our science and theirs are basically incompatible? We only have to consider the world 'view' of a blind person and a sighted person to realize that our perception of the world is very much dependent on sense data and how they are interpreted by the brain. In the case of a truly alien race (as opposed to a category of the human one), the difference in outlook might even be profound enough to prevent mutual recognition. If our scientific modes of communication do not overlap with theirs, it is possible that neither party will recognize the other: civilizations may exist oblivious of each other, like ships passing in the night.

Time is an unfortunate attribute of human mentality. This for a couple of reasons. First, different people view time in different ways, so it is a concept which inherently leads to confusion; and while the scientist may believe that he has a more concrete version of it, that version often does not mesh with the average person's everyday one. Second, humans are burdened with a sense of the passage of time, which causes a preoccupation with birth and death, of which the latter (at least) is a common source of angst. By contrast, animals appear to be unaware of time, living in the 'now'. They are unencumbered by the need to count up from a beginning or to count down to an end. The passage of time is a complicated mixture of biological and mental processes. That it is a curse can be appreciated by anybody who has observed the eager optimism of a child given a new toy, versus the blank pessimism of an old-aged pensioner staring into a glass of beer. We rationalize the burden of time by arguing that it is an unavoidable by-product of our higher-than-animal intelligence. This may be true, as far as it goes. But it could be that we as a race are stuck in a kind of intermediate valley of understanding; and that if we had a more advanced appreciation of time, we would come to see that events like birth and death are merely subjectively-identified points in an alternative matrix. Exactly what form this alternative may take is presently difficult to describe. However, a fresh gust of air into a musty subject is provided by the idea we outlined before: that time is a kind of ordering device. That is, the mind uses the concept of time to separate events which would otherwise overlap and be undecipherable, in the same way that we use the concept of spatial dimensions to organize the data we receive via our eyes and other senses. Time is, in short, probably an invention of the human mind.

Matter is the dominant concern of physics. For many people, the word "matter" still means "material", in the sense that it can be readily perceived by the human senses. The error of this perception was pointed out long ago by Eddington, who reminded us that what the senses regard as solid is in fact mostly empty space. And the history of the scientist's understanding of matter shows that it is a flexible concept, with no sharp division between "stuff" and "non-stuff" (vacuum). The history of the concept provides us with a strong hint that matter may not be purely a property of the world, but a convenient label provided by the mind. We emphasise, as before, that this viewpoint does not mean that matter does not exist. A person approaching a brick wall cannot *will* it out of existence, and anyone stupid enough to try would certainly end up with a bloody nose. What the viewpoint *does* say is that matter is an ephemeral thing which exists under certain conditions but not others; and that the physicist's understanding of these conditions is approaching the stage where the properties of matter can be deduced from purely mental considerations.

This is the epitome of mind weaving. That physics might be transformed into a kind of game played by highly-trained minds is a prospect that some workers find attractive and others find repellent. A lot could be written on both sides. However, it is undeniable that the subject has recently been moving in this direction. And below, we will present what we believe to be the first demonstration of how to go from an abstract thought to something equivalent to everyday matter, using only the thought processes of the human mind.

Experiments, by their nature, appear at first sight to run against the idea that physics can be a purely mental construct. Let us meet this criticism head-on, by considering some facts, which for convenience we arrange historically.

(a) The idea that the Earth goes around the Sun, rather than the other way around, was suggested by Greek philosophers including

Hipparchus long before the Polish astronomer Copernicus, who however rediscovered it and made it popular. The triumph of the one idea over the other had little to do with observations (which in historical times were scant and inaccurate), but a lot to do with the values of simplicity, or similarly of logic.

(b) The recognition by Newton, that the force which caused an apple to fall towards the Earth was also responsible for keeping the Moon in its orbit, was a colossal leap of intellectual faith. In the 1700s, such a conjecture must have appeared to be a flight of fancy.

(c) The paper by Einstein which proposed the special theory of relativity made no mention of the Michelson–Morley experiment; and while some controversy remains, Einstein appears to have been unaware and disinterested in experiments that related to the existence or otherwise of the all-pervasive fluid known as the aether (see Chapter 1). The sweeping-away of that medium, which had dominated work in physics for most of the 1800s, was basically a recognition of simplicity and a move towards logic by Einstein.

(d) The proposal a decade or so later of general relativity showed that Einstein's genius was no fluke. His field equations pointed the way to putting physics on a geometrical basis. However, even his stiletto-like mentality was not perfect (we will return to imperfections in mentality below). The mathematical structure of his equations suggested that he should add a term, which we nowadays relate to the cosmological constant — a move that was strongly endorsed by Eddington. At the time, Einstein assumed that the universe should be static. But when Hubble and others suggested that it was expanding,

Einstein dropped his cosmological constant. He would have been better advised to stick with his original logic, because modern models of the universe show that this term is in fact dominant.

(e) Particle physics, from its roots in the wave mechanics of the 1920s and 1930s, developed rapidly under the mathematical structures proposed by Schrodinger, Heisenberg and Dirac; and eventually settled into a regime described by quantum numbers, which are assigned on the basis of certain symmetry groups. That is, "real" particles are essentially described now by numbers associated with the properties of certain algebraic groups. (The latter are sometimes called internal groups, and are constructed along the lines of the Lorentz group of translations and rotations in external spacetime, which can be regarded as the analogous foundations of relativity.) The fact is that the particles which make up what people loosely refer to as "matter" are actually entities whose properties are completely described by mathematics.

(f) Coincident with the rise of particle physics, and somewhat overshadowed by it, Kaluza in 1920 and Klein in 1926 extended the 4D spacetime of Einstein's general relativity by adding an extra dimension. We have discussed this in detail elsewhere. It is not clear from their original papers if Kaluza and Klein regarded their fifth dimension as 'real' or as a mathematical abstraction. This ambiguity still bothers some modern physicists. But based on what we have discussed above, the distinction becomes in any event moot: if space and time are inventions of the human mind, then one (or more) extra dimensions are acceptable, provided they are introduced logically and lead to an improvement in understanding.

(g) Experiments such as the Large Hadron Collider cost a lot of money, and are regarded by some physicists as necessary for the advancement of the subject. However, there is a significant degree of opposition to such experiments by other physicists, who argue that even a fraction of the funds involved would support many young theorists, one of whom might be bright enough to answer the questions which the L.H.C. is designed to tackle. It is probably wise to decline to take a position in this controversy. But it is worth recalling that experiments are always designed with some theory as a basis, so they are not in any event entirely empirical in nature. (Even a simple synchrotron is designed assuming the validity of the laws of relativity, and would not operate if the latter were seriously in error.) The fact is that there is no sharp division between experiment and theory in science.

(h) Accordingly, the role of experiments in modern physics is largely one of the validation of theory, rather than what is naively regarded in some quarters as discovery. That said, the scepticism which some theorists show towards experiment may be unjustified. A nuts-and-bolts approach, while arguably crude, may still provide a short-cut to answering a physical problem.

(i) Astronomy occupies a position intermediate between the brute-force approach of the laboratory experiment and the ethereality of the purely theoretical calculation. It has its characteristic hardware, in the form of large and sophisticated telescopes, but their role is that of passive collectors of data. The data are quickly passed after acquisition to workers trained in analysis, who report on their findings and improve our understanding of the universe.

(j) The decline in the status of experimental physics has been long and insidious. To illustrate this, we mention the case of a recent article, which asked if experiments were not basically more important than theory in advancing the sciences. The paper was written by two noted physicists, from California and Canada, whose names we choose not to mention, because the article was rejected after peer review by several leading journals. Subsequently, one of the authors gave a talk to the physics department of a noted university, at the invitation of the present author. The talk attempted to catalog the various approaches to science through the whole of recorded history, ranging from the religious assumptions of olden times through the experiments of the Victorians to the theoretical jaunts of the modern era. The speaker ended by concluding that the latest models of cosmology, based as they are on pure thought, cannot be cataloged in a conventional manner. Indeed, the speaker (who was an experimental physicist) clearly regarded the flights of fancy of modern cosmology with disdain. By contrast, the audience (which consisted of both experimental and theoretical physicists) was uniformly of a different opinion, as shown by their questions and comments. The consensus view was that science is not so much dependent on practical considerations as it is on logical ones; and that at least today science is a subject of finely-honed theories and their mutual fit.

The preceding discussion of experiments is admittedly lengthy. But it is important to realize that there has been a century-long change in the attitudes of physicists, and that the centre of gravity of the subject now lies in theory. Yesterday's fancies have become today's staples. There has also occurred a kind of 'physication' of the other branches of science, by which we mean that an element of physics-based theory has been added to subjects which were previously concerned with the collection and cataloging of data. Thus, geology has morphed into geophysics, with the conversion of Wegener's idea of continental drift into an analytical science whose paradigm is plate tectonics. Astronomy has evolved into astrophysics, where Hubble's data on the redshifts of galaxies now form a tiny part of modern cosmology as based on general relativity. Biology has undergone a complete transformation, because its descriptive character gained a purpose when Darwin proposed the origin of species through national selection (Eldredge 2005). Following from this, genetics has grown from an amalgam of biology, chemistry and physics into a discipline in its own right, one whose potential is enormous.

The life sciences continue to evolve at a dizzying rate; and since it is inherent to them that Man is but one animal among many, it is likely that religion will continue to decay as an influence on science. There are, of course, physicists who take a fundamentalist standpoint about the scriptures, as we noted above. But these individuals are regarded by the majority as interesting outliers in the community of science and sociology. Even for those scientists who have a well-defined spirituality, it is usually the case nowadays that their personal faith and their public science are disconnected.

Attitudes about how science is done — as opposed to its material achievements — also evolve. It was common in the 1970s that university courses in philosophy included as standard subjects the views of science-observers such as Kuhn and Popper. By the 1990s, however, most practising physicists had come to disregard many of the philosophical statements about the nature of scientific thought, in favour of simpler attitudes based on pragmatism. Revolutions in physics *do* occur in some sense; but in contrast to exaggerated reports in the media, basic changes often take decades to accomplish, their impact smoothed by the time it necessarily takes to assess their validity. Tests of the acceptability of new ideas in science are also more practical than the idealistic choice between right and wrong. It may be appealing philosophy to say that a theory can only be disproved but not proved. However, the working physicist is quite willing to accept that a theory is good given a reasonable degree of testing. After all, if a theory has been tested by (say) 100 runs of an experiment, there is little point in carrying out the procedure for the 101st time (especially if there is significant cost involved). The physicist sensitive to semantics may prefer to write in a paper that a theory has been 'validated' in preference to 'proven'. But that physicist knows in his heart that the theory in question has really been 'proven', and would probably bet money on it.

Trust in science — and physics especially — is remarkably widespread. Even the person who is unversed in the laws of mechanics implicitly trusts them when he or she drives a car. In fact, people put their lives into the hands of physics on a daily basis. It is integrated into our existence at every level, from travelling in a plane to buying things at the local market. Because of its use below, let us consider for a moment the mundane situation where a person wanders around a market, picking up various things that are potential purchases. Hefting things in the hand is an excellent way of checking their desirability, whether it applies to a food item like an apple or an object such as a cannonball (a housewife might well not have much use for the latter, but it serves the purpose of illustration). Hand-eye coordination is well developed in the average person, and the human brain automatically estimates the density of the item concerned. Archimedes long ago figured out an objective way to estimate the densities of things by comparing them with the density of water. We formalize this today by using as a standard the density of water under controlled conditions: 1 gram per cubic centimetre. (We could use different units, but that will not be important for our subsequent argument.) By definition, objects lighter than this will float in water, while those heavier will sink. So obviously apples float while cannonballs sink; but we make the concept general and useful by noting the relative density of materials with respect to water. Thus iron has a density relative to water of about 8, while gold is very hefty with a value of nearly 20. At the opposite end of the scale, gases usually have relative densities very much less than 1. In fact their essential physics is better described by another parameter, the pressure, which however can be related to the density by an equation of state (see above and Chapter 5). Concentrating on the density, let us agree to quantify all values of this parameter by the symbol $\rho$.

Previously, a promise was made: To give an example which shows that physics is invented rather than discovered. We now proceed to fulfil this promise, using the common-or-garden quantity $\rho$ as the centre-point of the analysis.

The following discussion is kept as short as possible, because we have already assembled the associated physics in Chapter 5. The demonstration follows the path set out by Eddington, who argued that science contains subjective elements that necessarily follow because all of our data are filtered through the human senses (Eddington 1928, 1939; Kilmister and Tupper 1962; Kilmister 1994; Batten 1994; Price and French 2004; Halpern and Wesson 2006). We will make use of an extra dimension in addition to the four of spacetime, but this is in line with our earlier investigations which showed that dimensions are products of the human mind. Five-dimensional physics is now well understood, and a technical proof of the following argument is available (Wesson 2006 or 2008). Even so, certain symbols appear in the following which may appear abstract but are necessary to keep the presentation to a reasonable length. Referring back to the beginning of our considerations in Chapter 1, technical symbols are just shorthand for objects that we have already defined. They should be regarded as benign things, in the same manner as a picture sums up a wordy description. By way of an abstract: The argument begins with a tensor in a higher-dimensional space, and ends with the density of an apple in the market place.

The distance between two nearby points in a five-dimensional, imaginary space is given by an extension of the familiar formula due to Pythagoras. If the 'space' is curved, to describe forces like gravity, the potentials that correspond to the forces are encoded in a geometrical object called the Ricci tensor. In 5D, this is denoted $R_{AB}$, where the indices $A$, $B$ run over time, space and the extra dimension.

We can regard $R_{AB}$ as a $5 \times 5$ array, though the quantities on one side of the diagonal are equal to those on the other, so it really has only 15 independent components. The simplest field equations, which are also the 5D analogs of the 4D ones used to test general relativity in the solar system, are $R_{AB} = 0$. However, when matter is present, its properties are usually encoded in a $4 \times 4$ array, the energy-momentum tensor $T_{\alpha\beta}$, where $\alpha$, $\beta$ run only over time and space. This object has only 10 independent components. It gives back the laws of motion and the law of conservation of mass-energy by setting its divergence ('flow') to zero. Its 0-0 or time-time component gives the common-or-garden density $\rho$. Clearly, in giving a completely abstract description of the common density $\rho$ we need to go from the 5D object $R_{AB}$ to the 4D one $T_{\alpha\beta}$. Fortunately, Campbell's theorem shows us how to do this. The working necessary to go from 5D to 4D is what physicists call "tedious" (i.e., boring). It consists of decomposing the 15 equations of $R_{AB} = 0$ into sets of 10, 4 and 1. The last two sets give a wave equation for the new or scalar potential connected with matter, and a quartet of conservation laws for this. The main set of 10 equations turns out to be identical to Einstein's equations of general relativity, but with a definition for the energy-momentum tensor $T_{\alpha\beta}$ which is precise and based on the geometry of the extra dimension. This result was given previously as equation (5.6) of Chapter 5. That equation is somewhat cumbersome; but thought shows it has to be, in order to account for all possible forms of matter. That it does this has been shown by numerous applications. The 0-0 component of the 5D version of the 4D object is the common-or-garden density as measured by (say) a housewife at the market. To sum up: we have gone from an imagined higher-dimensional 'space' to the density as understood in everyday life. Q.E.D., or quod erat demonstrandum, meaning that we have proven what we set out to do.

The full proof of the above result takes several pages of tight algebra, but it has been widely studied by mathematical physicists since its appearance (Wesson and Ponce de Leon 1992). The result clearly has immense implications for philosophy: we start with an imaginary 'space', follow an intricate series of mathematical steps, and end up with something that is the same as what everybody understands by "density". (The calculation can be extended to include other properties of matter, such as pressure and heat flow, all of which match everyday experience.) It is remarkable in itself that we can follow a chain of pure thought and arrive at the same thing as what everybody "knows" as density, be it that of an apple or a cannonball. The implication is profound: physics is arguably invented, not discovered as often assumed.

It is of course, a step from the invention of the properties of matter as they are understood by physics to asserting the same thing about all of science. However, it is impossible to avoid the implication that science as a whole is a construction of the human mind.

An objection to this view is: Why, if it's a mental construct, are we not aware of all of physics (say) *now*?

An answer to this is the obvious and practical one: The typical human brain is not a perfect reasoning machine. Some people do not have any aptitude for the natural language of physics, namely mathematics. And even a professional mathematician, who can work swiftly and accurately later in the day, is unlikely to be as sharp just after getting out of bed. The same organ which is the source of wondrous inventions is also prone to stupid mistakes and wrong routes of reasoning. It is the case of the striving synapses: if things go right we get a masterpiece, but if they go wrong we get a mess.

It is a legacy of evolution that there is a wide variety in the capabilities of human brains. Some are able to create beautiful works of art, and some are tuned to produce insightful theories of physics; while others (the majority) are fully occupied with the humdrum demands of existence. Even a chosen brain has its highs and lows, depending on everything from banal events like the timing of the last meal to subtle psychological influences that can stimulate or depress the creative process. Creativity and emotivity are probably gene-linked, so that the imagination of the artist and the scientist is frequently a path through a minefield of moods. It can be argued that the happiest people in society are those who are not lumbered with the onus of creativity; though the large numbers of these people are indispensable, in forming the broad base of a kind of communal pyramid which supports creators at its apex. It is actually somewhat remarkable that scientists, as a group of creators, manage to produce a coherent body of useful knowledge. If we accept, for the moment, that scientists get their results from their own minds, we are almost obliged to ask how exactly they do this. That is: What is the precise mechanism whereby science is created; and how do its practitioners view it?

Chess is a fascinating pastime, and many physicists liken the practice of their science to the playing of this game. The analogy is instructive. Chess is an intricate game, whose rules are rigid but provide enough scope by their combination as to provide many alternatives that tax even the best minds. The parallel with physics is obvious, insofar as mathematics provides the rules, which can be combined in a formally infinite variety of ways. However, the analogy becomes inaccurate beyond this stage. For example, the foundations of mathematics remains a subject of controversy. Some believe that mathematics is based in the primitive concept of number (e.g. Peano), with others believe that it is based in the application of logic (e.g. Russell). Also, mathematics is not static like the rules of chess, but evolving with time; and while its new parts are required to be consistent with the old parts, several important developments have not been purely abstract, but the result of input from the practical demands of physics. (A notable case is provided by Dirac's delta function, which was introduced to describe certain physical distributions along an axis, but is only defined via its integral along that axis, thus providing the starting point for the theory of functionals as opposed to ordinary, defined-at-a-point functions.) Further, chess is after all only a game, whereas we live physics as an everyday experience. It is, nevertheless, interesting that some physicists feel as if they are playing a game when they manipulate their equations. As far as it goes, this attitude agrees with the thesis being proposed here, that physics and science in general are not so much discovered as invented.

Even if chess and physics were equivalent, there is still the question of where the knowledge ultimately comes from which we call science. Since we are nearing the end of our presentation, and are on new and precarious ground, we will address this issue in short form. Let us consider a few ideas about the origin of our science, presuming that the organ mainly concerned lies between a person's ears.

(i) The human mind may be far more powerful than previously acknowledged. It is often said that the brain appears to use only 10% of its capabilities in running the basic functions of a person's life. If this is true, with what is the other 90% occupied? It could be the storage, retrieval and analysis of more abstract things, such as science. We will return to this below.

(ii) A less likely option is that science is not "located" in any one brain, but is distributed among many. This goes back to the old idea of a shared racial memory. Humans certainly share many inherited mental traits; but it is difficult to see how the abstract notions necessary to account for a subject like relativity could be collected from disparate brains and integrated by one person like Einstein to produce a coherent account.

(iii) The many-worlds interpretation of quantum mechanics may offer in principle a better scheme, whereby the data content of many brains can be collated in one of them (see Chapter 2 for an outline of this approach to physics). Research is ongoing into non-local quantum field theory which may be relevant to this problem; but at present there is no known mechanism for gaining access to the science that may be encoded in many versions of reality and concentrating it into the experience of one or a few people.

(iv) God may also be a potential source of our science, particularly for those who believe in his all-pervasive existence. This is a religious version of the idea that underlies (ii) and (iii) above, in which we regard the human brain as a kind of focussing device for scientific information that is otherwise distributed in some large, mental realm. However, science is not correlated with religiosity (see Chapter 6), and certainly not the privilege of the pious.

(v) Solipsism is always a solution to any problem, in society or science. It is conceivable that all aspects of a person's experience are internal fantasy, including physics and the people we usually associate with it. However, even Eddington — who was the first notable scientist to admit subjectivity — was careful to note that he believed in the existence of an external world. Solipsism, almost by definition, cannot be refuted. But this in itself makes the option uninteresting, at least for the scientist.

The foregoing ways in which we can account for science as a mental product are not exhaustive, but they do serve to sample the range of possibilities. Although a philosopher might be willing to consider one of the other options, the average physicist would pick (i) as the most promising alternative. Let us therefore consider this more closely, and draw our deliberations to a close by making a few reasonable conjectures.

Brain power is the main factor which distinguishes humans from other animals. Paradoxically, however, we understand less about the brain than about any other organ in the body. This leaves a lot of latitude to speculate about its capabilities. One possible capability which we have mentioned, due to Penrose, is that the human brain may amplify microscopic events to macroscopic ones, thereby bridging the gap between quantum and classical physics (Penrose 1989; Abbot, Davies and Pati 2009). While conjectural, this idea has received serious attention. But it may only be one of a variety of things that the brain can do, of which we are presently ignorant.

We have presented evidence from several disciplines — and mainly physics — that science is not so much discovered as invented. That is, invented by the human mind. As such, science is akin to fine art, music and the other creations at which people excel. We create our culture as we go along. Artists know this; but many scientists have not seriously considered the option, preferring to believe that they are merely discovering things. This attitude, on reflection, is both odd and problematical. It is odd because it puts science at variance with the other achievements of humanity. And it is problematical, because if science is merely discovered, who or what was responsible for it? God or some other benign intelligence? Hardly. Science depends on inspiration and skill, and is a cousin to the Arts.

Given this interpretation, the human mind must be a far more inventive and imaginative organ than previously assumed. In a way, this is already apparent, because science-fiction writers create alternative realities which (at least in some cases) are quite plausible. An example is provided by the classic movie *2001: A Space Odyssey*. The brainchild of Stanley Kubrick and Arthur C. Clarke, the action is believable for a couple of hours because it lies on the border between the mundane and the magical. (Clarke once remarked that any sufficiently advanced technology would appear to be magical.) If such a movie is believable, we can certainly consider the possibility that the physics it portrays is also a creation of the mind.

No human mind is infallible, however; and some individuals are better than others in retrieving and enunciating science. That is why we respect the great scientists like Newton and Einstein. It is not an accident that we refer to such people as "thinkers". What causes some individuals to promulgate their science while others remain quiet is obscure. Darwin, among the greats, was refreshingly open about his motives: "I worked ... from the mere pleasure of investigation ... But I was also ambitious to take a fair place among scientific men." In other words, his urge to do science came from a mixture of pure interest and the wish to be recognized. That is OK. Probably most people would admit to a similar psychological combination, irrespective of the nature of their job. But while some jobs are simple and easily mastered, others are complicated and require meticulous practice. Research in the mathematical sciences is like this. As noted elsewhere, a theoretical physicist should no more make a mistake in his calculations than a concert pianist should play a bum note. However, most scientists know the craft of their chosen occupation very well. The dog-legged history of science is, on the basis of our present thesis, due to the imperfect way in which the brain accesses and processes its hidden stores of knowledge.

The human brain may be the biggest treasure chest imaginable of science. The mind — beneath its humdrum daily activities — may well give a person access to new and fascinating aspects of physics and related subjects. Exactly how this trove is opened depends on the individual. But once opened, the deep mind might allow of the creation of fresh and almost magical science.

## References

Abbot, D., Davies, P.C.W., Pati, A.K. (eds.), 2009. Quantum Aspects of Life. World Scientific, Singapore.

Batten, A., 1994. Quart. J. Roy. Astr. Soc. 35, 249.

Dawkins, R., 2006. The God Delusion. Bantam, New York.

Eddington, A.E., 1928. The Nature of the Physical World. Cambridge University Press, Cambridge.

Eddington, A.E., 1939. The Philosophy of Physical Science. Cambridge University Press, Cambridge.

Eldredge, N., 2005. Darwin: Discovering the Tree of Life. Norton, New York, p. 27.

Halpern, P., Wesson, P.S., 2006. Brave New Universe: Illuminating the Darkest Secrets of the Cosmos. J. Henry, Washington, D.C.

Kilmister, C.W., Tupper, B.O.J., 1962. Eddington's Statistical Theory. Clarendon Press, Oxford.

Kilmister, C.W., 1994. Eddington's Search for a Fundamental Theory: A Key to the Universe. Cambridge University Press, Cambridge.

Penrose, R., 1989. The Emperor's New Mind. Oxford University Press, Oxford.

Price, K., French, S. (eds.), 2004. Arthur Stanley Eddington: Interdisciplinary Perspectives. Centre for Research in the Arts, Humanities and Social Sciences (10–11 March), Cambridge.

Tipler, F.J., 2007. The Physics of Christianity, Doubleday, New York.

Wesson, P.S., Ponce de Leon, J., 1992. J. Math. Phys. <u>33</u>, 3883.

Wesson, P.S., 2006. Five-Dimensional Physics: Classical and Quantum Consequences of Kaluza–Klein Cosmology. World Scientific, Singapore.

Wesson, P.S., 2008. Gen. Rel. Grav. <u>40</u>, 1353.

# INDEX